Compost Utilization in Production of Horticultural Crops

Compost Utilization in Production of Horticultural Crops

Edited by
Dr. Monica Ozores-Hampton
CEO and Co-Founder of TerraNutri, LLC

CRC Press
Taylor & Francis Group
Boca Raton London New York

CRC Press is an imprint of the
Taylor & Francis Group, an **informa** business

First edition published 2021
by CRC Press
6000 Broken Sound Parkway NW, Suite 300, Boca Raton, FL 33487-2742

and by CRC Press
2 Park Square, Milton Park, Abingdon, Oxon, OX14 4RN

© 2021 Taylor & Francis Group, LLC

CRC Press is an imprint of Taylor & Francis Group, LLC

Library of Congress Cataloging-in-Publication Data

Names: Ozores-Hampton, Monica, editor.
Title: Compost utilization in production of horticultural crops / Monica Ozores-Hampton.
Description: First edition. | Boca Raton, FL : CRC Press, [2021] | Includes bibliographical references and index. | Summary: "Compost Utilization in Production of Horticultural Crops provides information for the compost industry to develop horticulture production efforts and techniques. This highly practical book contains information applicable to current production issues facing the fruit and nuts, vegetable, and ornamentals and turfgrass industry. Written by scientific experts, chapters evaluate the uses of compost for greater crop yields and decreased plant disease and pesticide application, irrigation water and fertilizer demand. Considering compost use increases carbon sequestration, the book provides guidelines on converting safe waste materials into composted soil amendments while minimizing negative impacts on the environment. Chapters cover the diversity and variability on compost uses of available feedstocks; composting methods, application rates, methods and timing; and considers the benefits of application alone or combined with other organic or inorganic nutrient sources"-- Provided by publisher.
Identifiers: LCCN 2020043159 (print) | LCCN 2020043160 (ebook) | ISBN 9780367691073 (paperback) | ISBN 9780815366461 (hardback) | ISBN 9781003140412 (ebook)
Subjects: LCSH: Compost. | Compost plants.
Classification: LCC S661 .C667 2021 (print) | LCC S661 (ebook) | DDC 631.8/75--dc23
LC record available at https://lccn.loc.gov/2020043159
LC ebook record available at https://lccn.loc.gov/2020043160

ISBN: 9780815366461 (hbk)
ISBN: 9780367691073 (pbk)
ISBN: 9781003140412 (ebk)

Typeset in Palatino
by Deanta Global Publishing Services, Chennai, India

Dedication

This book is dedicated to my parents Carlos and Lucia

*Who always encouraged me to learn and
pursue challenging goals in my life*

*The compost journey has been a wonderful experience full
of surprises and adventures so I am eternally grateful*

Contents

List of figures

List of tables

Editor biography

Monica Ozores-Hampton is a CEO and Co-Founder of TerraNutri, LLC. Dr. Ozores-Hampton obtained her BS in Horticulture from Universidad Católica de Chile, Chile; her MS in Biological Science from Florida International University, Miami, Florida; and her PhD in Horticultural Sciences from the University of Florida, Gainesville, Florida. She was a former Associate Professor of the Department of Horticultural Sciences at the University of Florida, Immokalee, Florida. She has been working in the compost and horticulture industry due to her knowledge, skills and experience, nationally and internationally. She is known as the "Compost Queen" by the industry. She specialized in nutrient management, plant-soil nutrient cycling and mineral nutrition in plant science and production. In the area of compost and composting, she has experience in designing, permitting, compliance inspections, reporting and operating composting facilities nationally and internationally. Additionally, she developed compost quality assurance/quality control and testing programs using many compost feedstocks in horticulture crops. She dedicated decades to research, teaching and extension work in compost production and utilization in horticulture crops. During her tenure at the University of Florida, her grants from external and internal sources total more than $1.5 million and $1 million as in-kind contributions from the horticulture industry. Among her accomplishments, including being a recipient of the Rufus Chaney compost researcher of the year award (US Composting Council (USCC)), she has received Award Young or Early Career Professional Leadership Symposium Award from the American Society for Horticultural Science. She also received the Excellence in Crop Production Award from the National Association of County Agricultural Agents and Council Memorial Tomato Research Award. She has been the editor of the Florida Tomato Institute Proceedings and consultant editor of journals such as *HortScience* and *Compost Science and Utilization*. She has authored or co-authored many book chapters, refereed research and extension publications. She has published 60 refereed research publications,

1 compost book (*In print*), 9 book chapters and 150 non-refereed publications. She has been an editor of the *Journal of Sustainable Agriculture*™, *HortScience* and *Compost Science and Utilization*. She has been regularly invited to speak at hundreds of conferences, to teach at the composting school of the USCC, to organize and participate in compost training and workshops, nationality and internationally.

Contributors

Monica Ozores-Hampton
TerraNutri, LLC

Fernando Alferez
University of Florida

Craig Coker
Coker Composting and Consulting

David Hill
CycleLogic

Rod Tyler
Green Horizons Environmental, LLC

chapter 1

Past, present and future of compost in horticulture crop production

Monica Ozores-Hampton

Contents

Composting is an aerobic biological decomposition process where microorganism's breakdown and convert raw organic materials (plant and animal) into relatively stable humus-like materials to improve soil chemical, biological and physical characteristics (Ozores et al., 2011; Ozores-Hampton, 2017a and b). Composting serves as both a waste management method and a product manufacturer: a compost producer can generate revenue at both the front end of the composting process such as at the beginning with tipping fee and at the back end by selling the compost (Ozores-Hampton, 2017a and b). The horticultural industry is the primary consumer of compost in the world. Therefore, the composting process represents the most widely used recycling technology of organic waste in agriculture (Stentiford and Sanchez-Monedero, 2016). Feedstock for composting can be generated from yard waste (YW), biosolids, municipal soil waste (MSW), animal manures (poultry, dairy, horse, swine and cattle with and without bedding) and other biodegradable waste by-products from urban or agricultural areas (Ozores-Hampton et al., 2011).

Proper composting can be expected to eliminate pathogens with high temperature (heat generation) to produce a stylized end-material, and therefore, no waiting period is required between application and crop harvest according to the Food and Drugs Administration Food Safety Modernization Act. An optimal composting process requires a selection of organic materials to obtain a favorable ratio of carbon (C) to nitrogen (N) of 25–30:1, adequate moisture (45–50%) and constant aeration (turning

Figure 1.1 R&D Soil Builders, Inc. Overview of windrow composting of yard trimming waste, Immokalee, FL. Credit: R. Robbins, R&D Soil Builders, Inc.

equipment). In the most common composting process, the organic material passes to a grinder to reduce the particle size and set out in long piles or windrows (Figure 1.1). The piles must be turned periodically to aerate and to assure homogeneity, preferably when core the temperature less is than 140°F as indicated by a compost thermometer. In favorable conditions, the compost can be finished in 3–6 months.

The main reasons for employing the composting process are the following:

1. Composting is a more environmentally enhancing method of disposal of organic waste material compared to landfill or incineration.
2. The composting process can accomplished volume and weight reduction of 30%–60%. Therefore, tremendous labor and cost reduction of transportation and spreading of the end product.
3. Exposing materials to high composting temperature eliminates human, plants pathogens, nematodes and weeds seeds.
4. Manures, biosolids and other wastes are attractive to flies, other insects or animals that lay eggs which can be eliminated or minimized by the high temperature.
5. Materials can be blend and composted together, enhancing the composting process and final quality of the end product.

6. Organic waste materials need to be composted in a composting facility in which the composting process can undergo a manage control temperature instead of land or soil being used as a waste management tool.
7. Composting improves the soil quality by maintaining or increasing soil organic matter (SOM) content, providing slow-release nutrients compared to inorganic fertilizers and therefore pollution of water resources from nutrients is reduced.

Along with these advantages, composting has some disadvantages such as increased labor cost, transportation, knowledge and equipment. Implementation of the suboptimal composting practices can have negatives consequences during the composting process such as inadequate C:N ratio slowing the process or losses of N by volatilization or ultimate poor-quality compost that reduced crop growth, yield and quality.

In the United States, 267.8 million (tons) of MSW, 24.4 million tons of YW, 8 million tons (dry weight basis) of biosolids, 136 million tons of dairy cow, beef cattle, swine and poultry manure wastes (dry weight basis) and 39.7 million tons food waste were produced with a generation rate of 8.0 lb/person/day in 2017 (U.S. Environmental Protection (USEPA) Platt et al., 2014; USEPA, 2017). Compost produced from all the waste streams combined was 106.6 million tons with a rate of 2.1 lb/person/day (USEPA, 2015). Municipal solid waste is the largest category with a common destination of the waste being 52% landfill, 25% recycling, 13% energy recovery and 10% composting (Figure 1.2). However, a large portion of the potentially organic compostable materials in the MSW remains in the waste stream and has not been composted (Figure 1.3).

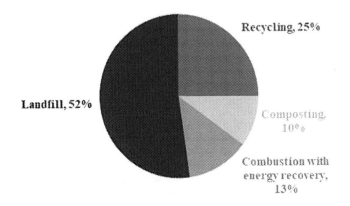

MUNICIPAL SOLID WASTE
MANAGEMENT 2017

Recycling, 25%

Landfill, 52%

Composting, 10%

Combustion with energy recovery, 13%

Figure 1.2 Destination of municipal solid waste in the United States during 2017.

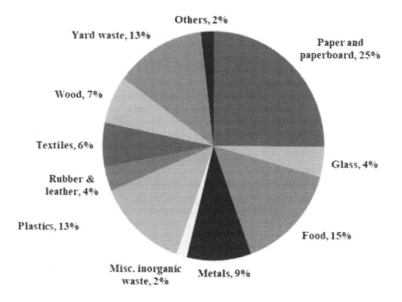

Figure 1.3 Municipal soil waste by categories during 2017.

Of the MSW generated, approx. 27 million tons or 35.2% were composted including approx. 24.4 million tons of YW and 2.6 million tons of food waste (USEPA, 2017). However, MSW composting rate was only 0.45 lb/ person/day. The rate of YW that has been composted is 69.4% and food waste is 6.3% (USEPA, 2017). In the United States, there are 4,914 composting facilities operating in 44 states which produce 70% YW, 9% MSW, 8% on-site on-farm/agricultural operation, 7% on-site institution, 5% BS and 1% other wastes (Platt and Goldstein, 2014). The top compost producing states are California, Florida, Iowa, Washington and New York; each of these states produced more than 1 million tons of compost per year (Platt et al., 2014; Platt and Goldstein, 2014).

1.1 History of compost and composting in horticulture production

Composting is among the oldest fertilizer practices to improve soil fertility with evidence of it been used by the Empire in the Mesopotamian Valley, Romans, Greeks and Tribes of Israel. In Mesopotamian, more than 10,000 years ago, compost was useful in the transition from a society of hunter-gatherers to localize their food production system in one place. Animal manure application provided the soil with the nutrients required to be able to produce food crops for human combustion. However, the first references of composting were from more than 6,000 years ago by the Chinese (Stentiford and Sanchez-Monedero, 2016). Limited information about composting and compost utilization in crops is available until Sir Howard, in the 20th century, introduced a composting method known as "the Indore process" based on his extensive research during his work in India between 1924 and 1931 (Hershey, 1992; Howard and Wad, 1931; Fitzpatrick et al., 2005). The Indore composting process was inspired by the Chinese in a need to provide Indian growers with inexpensive nutrients to grow vegetable crops since standard fertilizers were not cost-effective. The composting process consisted of a mix of several components: three parts of green materials to one part of animal manure in multiple layers to a height of 1.4 m (Howard, 1943). The Indore method had limited turning, and therefore multiple modifications in the pile turning were considered critical to accelerate the composting process significantly and were done throughout the years (Diaz and Bertoldi, 2007).

In the middle of the present century, two types of composting technology were developed to improve aeration (turning): the "windrow" and the Dano drum (Diaz and Bertoldi, 2007; Fitzpatrick et al., 2005). The windrow composting consisted in placing the mixture of raw materials in long narrow piles which are agitated or turned on a regular basis. The turning will rebuild the porosity, release trapped heat, water and gasses (CO_2, NH_3 and others) and exchange the materials at the windrow surface with materials from the interior where weed seeds, pathogens and fly larvae are destroyed by high temperature (Fitzpatrick et al., 2005). The Dano horizontal drum (2.7–3.7 m diameter) is a fast composting process of 3 to 7 days and is an expensive technology with effective pathogen destruction and therefore popular with the waste management industry obtaining tremendous volume reductions. However, MSW composting produced low-quality materials due to inert material such as glass contamination (Farell and Jones, 2009; Fitzpatrick et al., 2005). In 1970, a new technology was developed in the United States called the "static pile composting method" which eliminates the need for turning by supplying positive air or removing negative air by blowers to/from the composting materials through perforated pipes embedded in each windrow (hot gases rise upward out of the windrow). Therefore, no turning or agitation of the feedstock materials occurred once the pile was formed at the composting site. The most popular materials used with the static pile composting method were biosolids and wood chips, which allowed control of the temperature and odors (Raviv, 2005; Fitzpatrick et al., 2005).

1.2   Compost use in horticultural crops

Compost can be used in horticulture including food and ornamental and turfgrass crops production. Food crops include vegetables, fruits and nuts. Most vegetable crops are grown annually, while fruits and nuts are perennial (one or more years). Ornamental crops include turfgrass production and field nursery crops (trees and shrubs) (Ozores-Hampton., 2012). The main considerations for successful compost use in annual or perennial production systems are included in Figure 1.4. Parameters such as soil type, climatic conditions, crop type and compost attributes/rate/ application methods will determine the success in the production system in response to the compost application.

Compost can directly affect soil bulk density, water holding capacity, soil structure, soil carbon content, macronutrients and micronutrients, pH, soluble salts and cation exchange capacity and biological properties (microbial biomass) (Ozores-Hampton, 2012). Compost can be used in conventional and organic fruit (including nuts) and vegetable (annual or perennial) and ornamental crops (including turfgrass) production. Growers can use compost as a soil conditioner or as a nutrient source to supplement the fertility program in horticultural production (Rosen and Bierman, 2005). Nutrients such as N, phosphorous (P), and potassium (K) may be low content. To lower the environmental impact of high compost application rates and protect water supplies from excessive nutrient runoff or leaching or excessive soil nutrient buildup, compost should be applied to match the nutrient needs of a crop (Ozores-Hampton, 2017a and b). However, the compost quality use guidelines for assessing compost quality for use in fruit, vegetable and ornamental crop production are

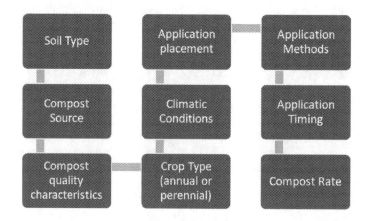

Figure 1.4 Compost use considerations in horticultural crops annual or perennial production systems.

limited to the end user. Nevertheless, the use of appropriate compost will improve soil quality and enhance the use of fertilizer, thereby improving the performance of crop production systems. Compost is a dynamic system; therefore, making recommendations for its use is more complicated than a standard fertilizer. The lack of knowledge of horticultural professionals' results in failure to use composts or mistakes and problems with compost production and use.

1.3   Compost future trends

Conventional crop fertility management is a whole-system approach intended to optimize on-farm resource cycling. Compost producers can use on-farm or locally available waste products from off the farm to economically supplement soil fertility, thereby creating a comprehensive fertility program that will build or maintain SOM content, which is associated with the soil's water and nutrient holding capacity and buffers against pH and temperature fluctuations. Matching crop nutrient requirements can be a challenging aspect in building the fertility program. Under-fertilization can reduce crop yield or quality. Overfertilization is inefficient and expensive and may contribute to nutrient runoff, groundwater pollution, soil toxicity, pest and disease susceptibility, excessive production of foliage and reduced crop quality. Therefore, future compost utilization in horticulture crops may focus on combining the organic and inorganic nutrient sources to match crop nutrient requirements during the critical portion or the whole season. The goal will be to increase crop yield, reduce leaching, improve nutrient use efficiency and reduce environmental impacts. Determination of N mineralization rates and improved understanding of how various composts affect P cycling will be a critical component of utilizing diverse compost feedstocks since mineralization rates will be affected by the interaction of compost type, soil, application rates and methods and environmental conditions. Accumulation of soil P needs to be a major priority in impaired watersheds, since P availability at different pH levels can increase the risk of eutrophication, which is the excessive growth of aquatic plants and algae caused by the addition of nutrients and sediments to water.

Advances in molecular microbial ecology are providing new tools available to investigate soil-plant health interactions, disease suppression and pathogen risk assessments in a range of compost-crop application scenarios. These advanced techniques can help determine the role of certain groups of microorganism in the soil associated with increase in yield and fruit quality or determine whether pathogens are present that can cause human or plant injury once applied to the soil, which can be critical for ensuring the success of compost utilization for crop production.

References

Diaz, L.F., and M. de Bertoldi. 2007. History of composting. In L.F. Diaz, M. de Bertoldi, and W. Bidlingmaier (eds.), *Compost Science and Technology, Waste Management*, Series 8, Elsevier, Oxford, UK, pp. 7–24.

Farrell, M., and D.L. Jones. 2009. Critical evaluation of municipal solid waste composting and potential compost markets. *Bioresource Technol.* 100(19):4301–4310. https://doi.org/10.1016/j.biortech.2009.04.029.

Fitzpatrick, G.E., E. Worden, and W. Vendrame. 2005. Historical development of composting technology during the 20th century. *HortTechnology* 15:148–151.

Hershey, D.R. 1992. Sir Albert Howard and the Indore process. *HortTechnology* 2:267–269.

Howard, A. 1943. *An Agricultural Testament*, Oxford University Press, London, UK, p. 85.

Howard, A., and Y.D. Wad. 1931. *The Waste Products of Agriculture*, Oxford University Press, London, UK, p. 35.

Ozores-Hampton, M. 2012. Developing a vegetable fertility program using organic amendments and inorganic fertilizers. *HortTechnology* 22:743–750.

Ozores-Hampton, M. 2017a. Past, present, and future of compost utilization in horticulture. *Acta Horticulturae*. International symposium on growing media, soilless cultivation, and compost utilization in horticulture. 27 Nov. 2020 *https://www.actahort.org/books/1266/1266_43.htm.*

Ozores-Hampton, M. 2017b. Impact of soil health and organic nutrient management on vegetable yield and quality. *HortTechnology* 27:162–165.

Ozores-Hampton, M., P. Roberts, and P.A. Stansly. 2012. Organic pepper production, pp. 165–174. In V. Russo (ed.), *Peppers: Botany, Production and Uses*, CABI, Cambridge, MA.

Ozores-Hampton, M., P.A. Stansly, and T. Salame. 2011. Soil chemical, physical, and biological properties of a sandy soil subjected to long-term organic amendments. *J. Sustainable Agri.* 35:243–259.

Platt, B., and N. Goldstein. 2014. State of composting in the U.S.A. *BioCycle* 55(6):19.

Platt, B., C. Cocker, and S. Brown. 2014. State of composting in the US. Institute for Local Self Reliance, Washington DC, USA. 28 May 2020. http://ilsr.org/wp-content/uploads/2014/07/state-of-composting-in-us.pdf.

Raviv, M. 2005. Production of high-quality composts for horticultural purposes: a mini-review. *HortTechnology* 15:52–57.

Rosen, C.J., and P.M. Bierman. 2005. Using manure and compost as nutrient sources for fruit and vegetable crops. Univ. Minnesota Ext. Serv. 28 May 2020. https://conservancy.umn.edu/handle/11299/200639

Stentiford, E., and M.A. Sanchez-Monedero. 2016. Past, present and future of composting research. *Acta Hortic.* 1146:1–10 http://dx.doi.org/10.17660/ActaHortic.2016.1146.1.

United States Environmental Protection Agency (USEPA). 2015. Advanced sustainable materials management: Facts and figures 2013: Assessing trends in material generation, recycling and disposal in the United States, Washington, DC. 28 May 2020. http://www.epa.gov/epawaste/nonhaz/municipal/pubs/2013_advncng_smm_rpt.pdf

United States Environmental Protection Agency (USEPA). 2017. National overview: Facts and figures on materials, waste and recycling. 6 March 2020. https://www.epa.gov/facts-and-figures-about-materials-waste-and-recycling/national-overview-facts-and-figures-materials.

chapter 2

Impact of compost on soil health

Monica Ozores-Hampton

Contents

The horticulture industry is familiar with production practices aimed to improve the health and function of the soil (Ozores-Hampton, 2017b). Arshad et al. (1996) defined soil quality or health as the continued capacity of the soil to function as a vital living ecosystem that sustains plants, animals and humans. The definition implies the management of the soil to provide food, shelter and water for future generations (Ozores-Hampton et al., 2011). The success of long-term horticulture production and maintenance of environmental quality is dependent on soil quality (Arshad et al., 1996). The decline in soil health can be aggravated by crop nutrient removal, leaching, decreased soil organic matter (SOM), application of inadequate outside inputs and intensive agricultural management (Agegnehu et al., 2015; D'Hose et al., 2014). The most common indicators of soil health or quality include measuring physical, chemical and biological soil properties (Doran and Parkin, 1994). Specific measurable indicators of soil quality include SOM, bulk density (BD), water holding capacity (WHC), cation exchange capacity (CEC), pH and soil microbial respiration (Arshad et al., 1996; Doran and Parkin, 1994).

The improvement of soil health is non-regulatory and incentive-based cultural practices that intend to reduce or prevent nutrient loss to the watersheds while maintaining horticulture productivity and profitability (Ozores-Hampton, 2012; Li et al., 2010; Stoffella et al., 2014; Ozores-Hampton, 2017a). The interactions among the soil physical, chemical and biological components can play a fundamental role in the sustainability and efficacy of land management in horticulture production (Gülser et al., 2015a,b; Montemurro et al., 2005; Pane et al., 2015). However, programs to minimize the movement of nutrients out of the root zone and reduce the resulting environmental impact can be challenging for the horticulture industry. Positive impacts in soil health may include the incorporation of compost, cover crops, raw manures and other soil amendments. These practices increase SOM, which can directly impact soil physical, chemical and biological properties (Center for Integrated Agricultural Systems (CIAS), 2002). Soil application of mature and stable compost may improve the ability of a plant to tolerate stress by slowing the release of nutrients (Ozores-Hampton et al., 2000; Hernando et al., 1989); adding SOM (McConnell et al., 1993); increasing available water content (AWC) and CEC (Serra-Wittling et al., 1996; Tester, 1990); decreasing BD (McConnell et al., 1993); decreasing erosion by water and wind (Tyler, 2001); increasing pH in acid soils (Hernando et al., 1989); and increasing soil microbial activity (Perucci, 1992). Microorganisms play a significant role in decomposition of SOM, which leads to the formation of humus and available plant nutrients. Finally, compost may reduce the levels of organisms that cause stress to plants, such as plant-parasitic nematodes and soil pathogens (McSorley et al., 1997; Hointink and Fahy, 1986).

In areas of high population, there are a variety of nonhazardous wastes generated for which composting and land application can provide an economically sound and environmentally acceptable option for utilization, but the majority of these wastes are currently landfilled or burned (Goldstein and Madtes, 2001; Ozores-Hampton et al., 1998). Compost can be produced from wastes generated by urban populations and can include municipal solid waste (MSW); yard waste (YW); food wastes from restaurants, grocery stores and institutions; wood wastes from construction and/or demolition; wastewater (from water treatment plants); and biosolids (BS). Agriculture produces other organic wastes that can be land applied or composted: poultry, dairy, horse, feedlot and swine manures; wastes from food processing plants; spoiled feeds; and harvest wastes (Ozores-Hampton et al., 1998; Ozores-Hampton, 2006; Ozores-Hampton et al., 2005).

Composted waste materials that are land applied can be incorporated as a soil amendment or as a source of nutrients for fruit trees, vegetables and nursery crops; they can be used to replace soil removed with nursery trees and sod or used as all or part of potting media (Ozores-Hampton, 2006; Ozores-Hampton and Peach, 2002). However, long-term application of compost made from waste materials must pass the federal and state regulations to be suitable for producing fruit and vegetable crops for human consumption (U.S. Environmental Protection Agency (USEPA), 1994 and 1999, 1995; FDEP, 1989).

2.1 Impacts of compost on soil physical properties

The utilization of mature and stable animal- or plant-based compost can improve soil physical properties including soil aeration, pore size

distribution, aggregate stability (AS; coarse-textured soils), soil structure, total porosity (TP), decreased bulk density (BD), penetration resistance, WHC, AWC, water infiltration rate (WIR), water aggregate stability
(WAS), saturated hydraulic conductivity (SHC), water drainage, lower surface runoff, soil erosion and surface crusting (Stoffella et al., 2014; Ozores-
Hampton et al., 1998, 2011) (see Table 2.1).

The compost is high in OM content (more than 25%), and its addition to the soil increases the SOM (Ozores-Hampton et al., 1998, 2011).
It is reported that high soil BD impaired functions such as restricted
root growth and poor movement of air and water through the soil (U.S.
Department of Agriculture (USDA), 2008). Long-term application of multiple feedstock compost (MSW, YW and BS) reduced soil BD compared with
no compost application (Ozores-Hampton et al., 1998, 2011). Arshad et al.
(1996) reported similar results, with compost application decreasing the
BD of the soil to an ideal range of less than 1.6 g/cm^3 or long-term application of YW compost reduced the BD from 1.21 g/cm^{-3} to 0.91 g/cm^{-3}.
Bulk densities above thresholds of 1.8 g/cm^3 in sandy soil are an indicator of low soil porosity and compaction (U.S. Department of Agriculture
(USDA), 2008). Compaction under humid conditions can result in shallow plant rooting and poor plant growth, influencing crop yield. Also, BD
can provide useful information in assessing the potential for leaching of
nutrients and erosion. Runoff and erosion losses of soil and nutrients can
be caused by excessive BD when surface water is restricted from moving
through the soil (Evanylo and McGuinn, 2000). When composted BS and
beef manure were applied to soils, the BD was reduced significantly as
compared to inorganic fertilizer application (Tester, 1990).

Studies in Florida with sandy soils using MSW compost increased
the AWC and WHC significantly, creating substantial water use efficiency
(McConnell et al., 1993; Gallaher and McSorley, 1994). The compost significantly increased soil moisture at field capacity between −8 kPa and −30
kPa by 35% (Ozores-Hampton et al., 2011). In general, OM physically holds
more water than mineral soil components such as sand, clay and silt; thus,
increasing the SOM content increases its WHC and AWC (Evanylo and
McGuinn, 2000). Similar results were reported in a soil quality study when
the researchers compared composted cotton gin trash and inorganic synthetic fertilizer application; the composted material in this case was a very
stabilized organic material which increased field moisture retention by
50% when compared to inorganic fertilizer (Evanylo and McGuinn, 2000).
Correspondingly, MSW compost applications at 146 tons/acre increased
the soil WHC and AWC by 43% (McConnell et al., 1993). In sandy soils,
the AWC surges are by increasing SOM content and the improvement
of the pore size distribution (Shiralipour, 1998). Application of compost
improves the pore size distribution by increasing the storage pores and
consequently increasing the WHC. The increase in WHC is associated

Table 2.1 The effect of compost feedstock source and rate on physical, chemical and biological properties of soils in open field and production

Compost feedstock[x]	Rate (tons/acre)	Effect[y]	Reference
		Physical properties	
MSW	0, 60 and 120	Increased SHC, WAS, AWC and decreased BD	Albaladejo et al., 2008
PMBS	0, 22.3, 44.6 and 66.9	Increased WIR, SHC, AS and AWC and decreased BD	Price and Voroney, 2007
MSW/BS	0, 26.8, 53.5 and 107	Increased SHC, WHC, TP, pore size distribution and AS and decreased BD	Aggelides and Londra, 2000
BS	0, 17.8, 35.7, 71.4 and 142.7	Increased AWC and TP and decreased BD	Navas et al., 1998
BS/MSW	0, 1.8 or 7.6	No effect on WAS	Debosz et al., 2002
Animal manure	0 and 30	Increased aggregate stability and BD	Gilser et al., 2015a
YW/animal manure	11 and 11/4.5	Increased TP, SHC and AWC and decreased BD	Celik et al., 2004
MSW	0, 4.4 and 8.8	Increased WAS and AS	Ferreras et al., 2006
MSW	0, 17.8, 35.7 and 53.5	Increased structural stability and WIR	Bouajila and Sanaa, 2011
YW/BS/MSW	0, 7.6, 12, 20 and 80	Increased WHC and decreased BD	Ozores-Hampton et al., 2011
BS	0, 2.2, 4.5 and 8.9	Increased AS and no effect on the WAS and BD	Guerrero et al., 2000
BS	0, 13.4, 26.8 and 40	Increased WIR, accumulated infiltration and WHC	Salazar et al., 2012
MSW	0, 67 and 115	Increased AS, AWC and and lower BD	Albaladejo et al., 2009
		Chemical properties	
MSW	0, 60 and 116	Increased SOC, TOC and humin (non-extractable Na₃PO₄); no effect on humic substances	Albaladejo et al., 2008
BS/MSW/YW	0, 7.1, 10.7 and 21.4	Decreased Zn and Pb but increased %total N, Ca and Mg and increased Cd to 10.7 tons/acre	Ozores-Hampton et al., 1994a

(*Continued*)

Table 2.1 (Continued) The effect of compost feedstock source and rate on physical, chemical and biological properties of soils in open field and production

Compost feedstock[x]	Rate (tons/acre)	Effect[y]	Reference
Farm compost	18, 36, 54 and 72	Increased pH, EC, total N and SOM	Chang et al., 2009
BS/MSW	0, 1.8 or 7.6	Increased resin-extractable Pi and mineralizable N	Debosz et al., 2002
PMBS	0, 22.3, 44.6 and 66.9	No effect on soil pH and EC and total N; increased soil C, C/N ratio, Cu and Co; decreased Cr and Se concentration	Price and Voroney, 2007
Poultry manure/TW	0, 13.8 and 64.2	Increased SOM, total N, P and NO_3 leaching	Evanylo et al., 2008
MSW/BS	0, 26.8, 53.5 and 107	Increased SOM, pH and CEC	Aggelides and Londra, 2000
MSW	0, 3, 6, 9 and 12	Increased total N and C, available P, K, Na and EC and no effect on heavy metals	Albaladejo et al., 1994
MSW	0, 4.4 and 8.8	Increased SOC	Ferreras et al., 2006
CGT/ST/YW	4, 9, 12.5 and 15	Increased Ca, K, Mg, Mn, SOM, total C and CEC	Bulluck et al., 2002
MSW	0, 17.8, 35.7 and 53.5	Increased SOC and total N	Bouajila and Sanaa, 2011
BS/MSW	0, 1.8 or 7.6	Increased mineralizable N	Debosz et al., 2002
YW/BS/MSW	0, 7.6, 12, 20 and 80	Increased P, K, Ca, Mg, Mn, Zn, SOM and CEC	Ozores-Hampton et al., 2011
BS/DM/YW	0, 36, 12, 17 and 20	No effect on Cd, Cu, Pb, and Ni on pepper fruits	Ozores-Hampton et al., 2005b
Animal manure	0 and 30	Increased SOC	Gülser et al., 2015a
MSW	0 and 22.3	Increased TOC and Fe and Pb, Cu, Mn and Zn	Madejón et al., 2001
PMBS/PM	6.2, 12.5 and 18.7	Increased soil pH, soil inorganic N and extractable P, K and Mg; no effect on heavy metal contents except increased or decreased Mn and Zn	Baziramakenga et al., 2001

(Continued)

Table 2.1 (Continued) The effect of compost feedstock source and rate on physical, chemical and biological properties of soils in open field and production

Compost feedstock[x]	Rate (tons/acre)	Effect	Reference
DM	0, 10, 20 and 40	Increased P and pH; no effect on K, Mg, Ca, Mn, Zn, S, NO_3, Mn, Na, K and Fe	Roe and Gerald, 2010
BS	0, 13.4, 26.8 and 40	Increased SOC, pH and P	Salazar et al., 2012
MSW	0, 67 and 111.5	Increased SOC; total N, P, K, EC and pH; and Cd, Zn and Cr, but no effect on Pb and Ni	Albaladejo et al., 2009
Biological properties			
MSW	0, 4.4 and 8.8	Increased soil microbial respiration	Ferreras et al., 2006
CGT/ST/YW	4, 9, 12.5 and 15	Increased beneficial microorganisms and decreased numbers of plant pathogens	Bulluck et al., 2002
MSW	0, 17.8, 35.7 and 53.5	Increased emission of $C-CO_2$	Bouajila and Sanaa, 2011
Farm compost	18, 36, 54 and 72	Increased soil microbial activity and enzymatic activity	Chang et al., 2009
YW/BS/MSW	0, 7.6, 12, 20 and 80	Increased soil microbial activity (species number and diversity)	Ozores-Hampton et al., 2011
PMBS/SWM	0, 5, 10 and 15.6	Increased enzymatic activities and MBC; no effect on CO_2-C	Lalande et al., 2003
MSW	0 and 22.3	Increased enzymatic activities and decreased organic phosphatase activity	Madejón et al., 2001
PMBS/PM	6.2, 12.5 and 18.7	Increased phosphatase activities, soil alkaline phosphatase activity and acid phosphate	Baziramakenga et al., 2001

(Continued)

Table 2.1 (Continued) The effect of compost feedstock source and rate on physical, chemical and biological properties of soils in open field and production

Compost feedstock[x]	Rate (tons/acre)	Effect[y]	Reference
BS	0, 2.2, 4.5 and 8.9	Increased viable fungal propagules and total number of bacterial cells forming units	Guerrero et al., 2000
MSW/DM	8.9 and 35.7 or 8.9	Increased MBC and microbial metabolism, but decreased ratio of MBC to SOC	García-Gil et al., 2000
BS	0, 13.4, 26.8 and 40	Increased biological activity: evolution of $C-CO_2$	Salazar et al., 2012
MSW	0, 67 and 115	Increased MBC, ATP content, dehydrogenase activity and soil microbial respiration	Albaladejo et al., 2009
MSW	0, 17.8, 35.7 and 53.5	Increased MBC and N; SOM and microbial diversification	Bouzaiane et al., 2007

[x] ATP = adenosine triphosphate; BS = biosolids; CGT = cotton gin trash; DM = dairy manure; MSW = municipal solid waste; PM = poultry manure; PMBS = paper mill biosolids; SD = sawdust; SM = sheep manure; ST = straw; SWM = swine manure; YW = yard waste.

[y] AS = aggregate stability; AWC = available water content; BD = bulk density; C = carbon; $C-CO_2$ = organic CO_2; Ca = calcium; Cd = cadmium; CEC = cation exchange capacity; Co = cobalt; Cu = copper; Cr = chromium; EC = electric conductivity; Fe = iron; K = potassium; Mg = magnesium; Mn = manganese; Na = sodium; MBC = microbial biomass carbon; N = nitrogen; Ni = nickel; NO_3 = nitrate; SOC = soil organic carbon; SOM = soil organic matter; P = phosphorous; Pb = lead; Se = selenium; SHC = saturated hydraulic conductivity; S = sulfur; TOC = total organic carbon; TP = total porosity; WIR = water infiltration rate; WAS = wet aggregate stability; WHC = water holding capacity; Zn = zinc.

with plant AWC, increased infiltration and reduced evapotranspiration. For horticulture producers, the easiest way to modify soil physical properties, especially in sandy soils, is by the addition of mature and stable compost or other organic amendments using appropriate soil management techniques.

2.2   Effect of compost soil chemical property

Mature and stable compost utilization can increase soil chemical properties such as soil organic carbon (SOC), total organic carbon (TOC), soil organic matter (SOM), cation exchange capacity (CEC), macronutrients (N, P, K, Ca, Mg and S), micronutrients (Cu, Co, Fe, Mn, Mo, Na, Se and Zn), heavy metals (Cd, Cr, Ni and Pb), C/N ratio and humic substances; however, the pH and electrical conductivity (EC) will depend on the feedstocks (Gülser et al., 2015a,b; Ozores et al., 2011) (see Table 2.1).

Applications of MSW compost showed improvement in SOC, TOC, SOM and humin substances (Albaladejo et al., 2008; Eldridge et al., 2014). Florida, long-term application studies of MSW, YW and BS to sandy soils has increased the level of SOM, pH, Mehlich 1-extractable P, K, Ca, Mg, Mn, and Zn, SOM and CEC and decreased soil pH (Clark et al., 2000; Ozores-Hampton et al., 2011). The SOM content increased more than 200% (threefold) in the first 12 inches, and the CEC of composted areas was about 2.5 times that of non-composted soils. Similarly, composted dairy manure application increased soil P and K contents, and compost applications of MSW or BS have significantly enhanced soil mineral N content (Johnson et al., 2006; Loper et al., 2010). Also, higher SOM and CEC content was obtained with the applications of 30 tons/acre per year of YW compost to an organic and conventional pepper (*Capsicum annuum* L.) farm in a 2 years study (Chellemi and Rosskopf, 2004). Correspondingly, Shiralipour (1998) reported that sandy soils application between 15 and 30 tons/acre of compost increased CEC by a minimum of 10%. The high level of EC was attributed to extensive decomposition of OM and increased nutrient concentrations. The increased nutrient concentration is a result of compost application, especially in BS and animal manures compared with plant-based compost. The effects of compost application in CEC depend on the loading rates (Shiralipour, 1998).

Application of MSW compost increases pH in acid soils (Hernando et al., 1989). Most compost has a near neutral or slightly alkaline pH with a high buffering capacity. The increase in pH in acid soils is due to the reduction or elimination of Al or Mn toxicity in pH below 5. The elevated calcium carbonate and basic pH in MSW compost are responsible for increasing the pH, creating a liming effect. Application of compost to neutral or calcareous soils does not change the pH.

Horticulture producers may apply compost as an organic N source and/or as a soil amendment (Ozores-Hampton, 2017a). Nitrogen, P and K content in compost are lower than in commercial synthetic fertilizers. Therefore, understanding the N mineralization rate of various compost feedstocks and synchronizing the compost application time and placement in the root zone with maximum plant uptake of N can be the optimal management of OM soil incorporation (Preusch et al., 2002). Compost N mineralization is the process by which microorganisms convert organic N into plant-available inorganic N by consecutive processes; first the organic N will be converted into NH_4-N and then into NO_3-N by the process of nitrification) (Diacono and Montemurro, 2010; Stoffella et al., 2014; Chazirakis et al., 2011; Habteselassie, et al., 2006). Generally, N mineralization of compost has a linear trend in the first stages of the processes, with a transitory and fast immobilization supported by soil microorganisms and then followed by net positive mineralization (Antoniadis, 2013; Baziramakenga et al., 2001; Cambardella et al., 2003; Hernández et al., 2014). The fate of the compost immobilization or mineralization will be determined by the amount of the C/N ratio of the composted material, initial availability of inorganic N, microbial activity, temperature and moisture content (Cambardella et al., 2003; Diacono and Montemurro, 2010; Flavel and Murphy, 2006; Hernández et al., 2014; Ozores-Hampton et al., 2012). The compost inorganic N content is the result of a dynamic exchange among the N released through microbial decomposition, microorganisms C requirement and the quantity of N integrated by the microbes feeding on the compost (Cambardella et al., 2003; Montemurro et al., 2005). The rate of conversion of soil organic N into plant-available forms of inorganic N will depend on the factors that affect the soil microorganisms (bacteria and fungi) activity, the source of composting materials, the degree of compost maturity, temperature, moisture content and soil quality (Deenik, 2006; Ozores-Hampton et al., 2012; Stoffella et al., 2014). However, the N availability is low, with only 10–15% of the total N in the first year and no residual effect in the second year (Shiralipour, 1998). Application of compost reduced nitrate leaching from agriculture soils. This is attributed to the slow release of N from the compost and consequent reduction in potential N losses via leaching as crop uptakes the soil N, thus reducing the amount of inorganic N fertilizers application and pollution of ground water.

Studies comparing traditional synthetic fertilizer application and 10 years of compost application in Canada indicate that soil C, Ca, Mg and Zn content was similar or higher in plots with compost application than in non-composted plots, but P and K soil content was higher in traditional fertilizer plots (Warman, 1998; Warman and Harvard, 1996). A 6-year study in an organically managed orange (*Citrus sinensis* L.) orchard showed higher total C and N content than in an inorganic fertilizer-managed orchard (Canali et al., 2004).

Another 7-year study on the application of MSW compost, BS and farmyard manure at three rates (11, 22 and 45 tons/acre) and one inorganic synthetic fertilizer in irrigated wheat–corn rotation showed increased CEC, SOC and OM (Hemmat et al., 2010). Application of manure or compost increased CEC of the soil by increasing the soil OM content and soil C content (Guibert, 1999). The effect was probably due to the increase in the fine fraction of the soil C content (Guibert, 1999). Normally, soil Al and Fe can fix P as complex and insoluble forms in acid soil conditions. The addition of OM neutralizes the reaction sites that could fix soil P. Organic lignin has an affinity for Al and Fe in the soil, inhibiting the soil from fixing P. The free P can be combined to form OM-P complexes, and then the P can be slowly released by microbial action into the soil (Brady, 1974). Even though sandy soils are less variable in their degree of compaction than finer texture soils, the surface layer (6 inches) is more subjected to compaction.

Among the diversity of compost commercially available, the P soil content can increase significantly to consecutive high compost applications of BS and animal manure and negatively impact the soil health and water quality. N, P and K ratio of most compost at the rate selected to satisfy N crop requirement will result in an excess level of P and an insufficient level of K. Therefore, application of synthetic fertilizer is required to bring the nutrient levels to balance crop requirement. Therefore, compost application on sensitive land to P addition should be done based on crop P rather than crop N requirements (Preusch et al., 2002; Sikora and Enkiri, 2003).

2.3   Effect of compost on soil biological property

Soil microorganisms are capable of decomposing complex C structures through specific enzymatic activities into simple organic and inorganic nutrients that could be available to grow horticulture crops (Pane et al., 2015). Application of mature and stable compost can increase soil biological properties in common soil quality indicators such as soil microbial respiration or activity, emission of $C\text{-}CO_2$, specific enzymatic activities, microbial biomass C (MBC), or count of beneficial microbes as species richness diversity (SRD) and total SRD (TSRD) or pathogenic microorganism (He et al., 2000; Ros et al., 2006; Table 2.1)

Compost application of MSW increased enzymatic activity and soil microbial respiration and decreased organic phosphate activity (Ferreras et al., 2006 and Madejón et al., 2001). Similarly, MSW, YT and BS compost application to sandy soil showed improvement in soil microbial activity as SRD and TSRD (Ozores-Hampton et al., 2011).

In the same study, the six soil functional groups (heterotrophic aerobic and anaerobic bacteria, N fixing bacteria, fungi, actinomycetes and pseudomonads) were increased by adding long-term mixed compost to the soil. The soil SRD was increased for four of the six functional groups tested. The largest increases were SRD for fungi and actinomycetes beneficial soil microorganisms. The overall TSRD of long-term mixed compost was significantly improved as compared to no compost application, indicating the beneficial effects of addition of C to sandy soils. Increased soil total microbial activity and TSRD by the addition of compost has been widely documented (Grobe, 1998; Serra-Wittling et al., 1996; Perucci, 1992; Press et al., 1996). The adoption of compost by the horticulture industry will have the potential to create pathogen suppressive soil by improving soil quality (Hointink et al., 1986, 1997). In general, soil-borne diseases develop in highly mineralized soils deficient in SOM (readily biodegradable OM) since these soils do not support the activity of microflora to suppress soil pathogens (Hointink et al., 1986, 1997). Actinomycetes are important in the nutrient cycling of chemical substances such as chitin and cellulose, improving soil structure and assisting in the reduction of plant pathogen pressures (Perucci, 1992). Pseudomonads are important in nutrient cycling, assisting plants with P availability and in the biological control of plant pathogens (Broadbent and Barker, 1974). The beneficial species of pseudomonads were higher and plant pathogenic oomycete fungi (*Pythium* and *Phytophthora* spp.) were lower in an organic production system than in chemically fertilized soils (Broadbent and Barker, 1974). These effects can be observed within a single season when compost is incorporated into the organic production system (Broadbent and Barker, 1974). Tree and vine crop growers were able to increase the soil fungal number and diversity by the application of YW compost and cardboard feedstocks, thereby enhancing nutrient uptake, disease suppression and drought tolerance (Grobe, 1998). These organisms are important for breaking down more complex organic compounds, stabilizing soil aggregates and controlling diseases (Grobe, 1998).

The goal for a horticulture industry and grower is to shift the microorganism community in the low-productive areas toward the same number and diversity present in high-yield soils (Hointink, et al., 1997). Studies of BS co-composted with MSW or YW soil application improved microbial soil activity by enhanced microbial biomass C and P. In the application of MSW compost, the increase in the microbial biomass C was due to the addition of the organic C (Bhattacharyya, 2003; He et al., 2000). However, BS compost may contain heavy metals such as Cu, Zn, Pb, Cd and Mn and Ni that may potentially reduce soil microbial activities and microbial biomass C of the soil (He et al., 2000; Stoffella et al., 2014).

Compost applications can improve physical, chemical and biological properties of soils by increasing SOM, C, pH, Mehlich 1-extractable P, K, Ca, Mg, Mn, Cu, Fe, and Zn concentrations, CEC, WHC, AWHC (available water holding capacity) and overall soil microbial activity, and by decreasing BD. Enhancing soil health with compost can recycle nutrients and potentially reduce nutrient leaching and environment nutrient impact. Additionally, the use of compost in horticulture production can improve yield and fruit quality.

References

Agegnehu, G.M., I. Bird, P.N. Nelson, and A.M. Bass. 2015. The ameliorating effects of biochar and compost on soil quality and plant growth on a Ferralso. *J. Soil Res.* 53:1–12.

Aggelides, S.M., and P.A. Londra. 2000. Effects of compost produced from town wastes and sewage sludge on the physical properties of a loamy and a clay soil. *J. Bioresource Technol.* 71:253–259.

Albaladejo, J., C. Garcia, A. Ruiz-Navarro, N. Garcia-Franco, and G.G. Barbera. 2009. Effects of organic composts on soil properties: comparative evaluation of source-separated and non-source-separated composts. Madrid, 12–13 November 2009. (1st Spanish Natl. Conf. on Adv. in Materials Recycling and Eco – Energy).

Albaladejo, J., J.C. Lopez, G. Boix-Fayos, G. Barbera, and M. Martinez-Mena. 2008. Long-term effect of a single application of organic refuse on carbon sequestration and soil physical properties. *J. Environ. Quality* 37:2093–2099.

Albaladejo, J., D.M. Stocking, and V. Castillo. 1994. Land rehabilitation by urban refuse amendments in a semi-arid environment: effect on soil chemical Properties. *J. Soil Technol.* 7:249–260.

Antoniadis, V. 2013. Mineralization of organic-amendment-derived nitrogen in two Mediterranean soils with different organic-matter contents, communications in soil science and plant analysis. *J. Commun. in Soil Sci. and Plant Analysis* 44:2788–2795.

Arshad, M.A., B. Lowery, and B. Grossman. 1996. Physical tests for monitoring soil quality, pp. 123–142. In J.W. Doran and A.J. Jones (eds.), *Methods for Assessing Soil Quality*. Soil Sci. Soc. Am. Spec. Publ. 49. SSSA, Madison, WI.

Baziramakenga, R., R.R. Simard, and R. Lalande. 2001. Effect of de-inking paper sludge compost application on soil chemical and biological properties. *J. Can. Soil Sci.* 81:561–575.

Bhattacharyya, P., K. Chakrabarti, and A. Chakraborty. 2003. Effect of MSW compost on microbiological and biochemical soil quality indicators. *J. Compost Sci. and Utilization* 11:220–227.

Bouajila, K. and M. Sanaa. 2011. Effects of organic amendments on soil physicochemical and biological properties. *J. Materials and Environ. Sci.* 2:485–490.

Bouzaiane, O., H. Cherif, N. Saidi, N. Jedidi, and A. Hassen. 2007. Effects of municipal solid waste compost application on microbial biomass of cultivated and non-cultivated soil in a semi-arid zone. *Waste Manage. Res.* 25(4):334–342.

Brady, N.C. 1974. *The Nature and Property of Soils*, MacMillan, New York.

Broadbent, P. and K.F. Baker. 1974. Association of bacteria with sporangim formation and breakdown in Phytophthora spp. *Aust. J. Agric. Res.* 25:139–145.

Bulluck, III., L. R., M. Brosius, G. K. Evanylo, and J. B. Ristaino. 2002. Organic and synthetic fertility amendments influence soil microbial, physical and chemical properties on organic and conventional farms. *J. Appl. Soil Ecol.* 19:147–160.

Cambardella, C.A., T.L. Richard, and A. Russell. 2003. Compost mineralization in soil as a function of composting process conditions. *J. European Soil Biol.* 39:117–127.

Canali, S., A. Trinchera, F. Intrigliolo, L. Pompili, L. Nisini, S. Mocali, and B. Torrisi. 2004. Effects of long-term application of compost and poultry manure on soil quality of citrus orchards in Southern Italy. *Biol. Fertil. Soils* 40:206–210.

Chang, E., R. Chung, and Y. Tsai. 2009. Effect of different application rates of organic fertilizer on soil enzyme activity and microbial population. *J. Soil Sci. and Plant Nutr.* 53:32–140.

Chazirakis, P., A. Giannis, E. Gidarakos, J.Y. Wang, and R. Stegmann. 2011. Application of sludge, organic solid wastes and yard trimmings in aerobic compost piles. *J. Global NEST* 13:405–411.

Celik, I., I. Ortas, and S. Kilic. 2004. Effects of compost, mycorrhiza, manure and fertilizer on some physical properties of a Chromoxerert soil. *J. Soil & Tillage Res.* 78:59–67.

Center for Integrated Agricultural Systems (CIAS). 2002. Building soil organic matter with organic amendments. College of Agricultural and Life Sciences, University of Wisconsin-Madison, 17 July 2015.

Chellemi, D., and E.N. Rosskopf. 2004. Yield potential and soil quality under alternative crop production practices for fresh market pepper. *Renewable Agric. Food Syst.* 19:168–175.

Clark, G.A., C.D. Stanley, and D.N. Maynard. 2000. Municipal solid waste compost (MSWC) as a soil amendment in irrigated vegetable production. *J. Amer. Soc. Agri. Eng.* 43:847–853.

Debosz, K., S.O. Petersen, L.K. Kure, and P. Ambus. 2002. Evaluating effects of sewage sludge and household compost on soil physical, chemical and microbiological properties. *J. Appl. Soil Ecol.* 19:237–248.

Deenik, J. 2006. Nitrogen mineralization potential in important agricultural soils of Hawaii. College of Tropical Agriculture and Human Resources (CTAHR). University of Hawaii at Mänoa, USA. 24 April 2020. http://www.ctahr.haw aii.edu/deenikj/Downloads/SCM-15.pdf.

D'Hose, T., M. Cougnon, A.D. Vliegher, B. Vandecasteele, N. Viaene, W. Cornelis E. Bockstaele, and D. Reheul. 2014. The positive relationship between soil quality and crop production: a case study on the effect of farm compost application. *J. Appl. Soil Ecol.* 75:189–198.

Diacono, M. and F. Montemurro. 2010. Long-term effects of organic amendments on soil fertility. A review. *Agron. Sustainable Dev.* 30(2):401–422.

Doran, J.W. and T.B. Parkin. 1994. Defining and assessing soil quality. In J.W. Doran, D.C. Coleman, D.F. Bezdicek, and B.A. Steward (ed.), *Defining Soil Quality for Sustainable Environment.* SSSA. Special Publ. No. 35. Am. Soc. Agron. and Soil Sci. Soc. Am., Madison, Wisconsin.

Eldridge, S.M., K.Y. Chan, N.J. Donovan, F. Saleh, D. Fahey, I. Meszaros, L. Muirhead, and I. Barchia. 2014. Changes in soil quality over five consecutive vegetable crops following the application of garden organics compost. *Proc. Ist IS on Organic Matter Management and Compost in Horticulture,* J. Biala et al. Acta Hort. 1018, ISHS:57–72.

Evanylo, G.K., C. Sherony, J. Spargo, and D. Starner. 2008. Soil and water environmental effects of fertilizer, manure and compost-based fertility practices in an organic vegetable cropping system. *Agric. Ecosyst. Environ.* 127(1–2):50–58.

Evanylo G. and R. McGuinn. 2000. Agricultural management practices and soil quality: measuring, assessing, and comparing laboratory and field test kit indicators of soil quality attributes. Virginia Cooperative extension Publication Number 452-400 21 May 2020 https://vtechworks.lib.vt.edu/bitstream/handle/10919/24675/VCE452_400_2000.pdf?sequence=1

Ferreras, L., E. Gomez, S. Toresani, I. Firpo, and R. Rotondo. 2006. Effect of organic amendments on some physical, chemical and biological properties in a horticultural soil. *J. Bioresource Technol.* 97:635–640.

Flavel, T.C. and D.V. Murphy. 2006. Carbon and nitrogen mineralization rates after application of organic amendments to soil. *J. Environ. Quality* 35:183–193.

Florida Department of Environmental Protection (FDEP). 1989. Criteria for the production and use of compost made from solid waste. Florida Administrative Code, Chapter 17-709. Tallahassee, FL.

Gallaher, R.N. and R. McSorley. 1994. Management of yard waste compost for soil amendment and corn yield, pp. 28–29. In *The Composting Council's Fifth Annual Conference. Proc.* Washington, DC. 16–18 November 1994.

Garcıa-Gil, J., C.C. Plaza, P. Soler-Rovira, and A. Polo. 2000. Long-term effects of municipal solid waste compost application on soil enzyme activities and microbial biomass. *J. Soil Biolo. Biochem.* 32:1907–1913.

Goldstein N. and C. Madtes. 2001. The state of garbage in America. 1999. *BioCycle* 42(12):42–52.

Grobe, K. 1998. Fine-tuning the soil food web. *BioCycle* 39(1):42–46.

Guerrero, C., I. Gómez, J.M. Solera, R. Moral, J.M. Beneyto, and M.T. Hernández. 2000. Effect of solid waste compost on microbiological and physical properties of a burnt forest soil in field experiments. *J. Biol. Fertility Soils* 32:410–414.

Guibert, H. 1999. Carbon content in soil particle size and consequence on cation exchange capacity of Alfisols. *Commun. Soil Sci. Plant Analysis* 30:2521–2537.

Gülser, C.F. Candemir, Y. Kanel, and S. Demirkaya. 2015a. Effect of manure on organic carbon content and fractal dimensions of aggregates. *Eurasian J. Soil Sci.* 4:1–5.

Gülser, C., R. Kızılkay, T. Askın, and I. Ekberli. 2015b. Changes in soil quality by compost and hazelnut husk applications in a hazelnut orchard. *J. Compost Sci. Utilization* 23: 135–141.

Habteselassie, M.Y., J.M. Stark, B.E. Miller, S.G. Thacker, and J.M. Norton. 2006. Gross nitrogen transformations in an agricultural soil after repeated dairy-waste application. *J. Soil Sci. Soc. Amer.* 70:1338–1348.

He, Z.L., A.K. Alva, D.V. Calvert, Y.C. Li, P.J. Stoffella, and D.J. Banks. 2000. Nutrient availability and changes in microbial biomass of organic amendments during field incubation. *J. Compost Sci. & Utilization* 8:293–302.

Hemmat, A., N. Aghilinategh, Y. Rezaineja, and M. Sadeghi. 2010. Long-term impacts of municipal solid waste compost, sewage sludge and farmyard manure application on organic carbon, bulk density and consistency limits of a calcareous soil in central Iran. *Soil Tillage Res.* 2:43–50.

Hernández, T., C. Chocano, J.L. Moreno, and C. García. 2014. Towards a more sustainable fertilization: combined use of compost and inorganic fertilization for tomato cultivation. *J. Agri. Ecosyst. Environ.* 196:178–184.

Hernando, S., M.C. Lobo, and A. Polo. 1989. Effect of the application of municipal refuse compost on the physical and chemical properties of a soil. *Sci. Total Environ.* 81:589–596.

Hointink, H.A.J., and P.C. Fahy. 1986. Basis for the control of soilborne plant pathogens with composts. *Ann. Rev. Phytopathol.* 24:93–144.

Hoitink, H.A.J., A.G. Stone, and D.Y. Han. 1997. Suppression of plant diseases by composts. *HortScience* 32:184–187.

Johnson, G.A., J.G. Davis, Y.L. Qian, and K.C. Doesken. 2006. Topdressing turf with composted manure improves soil quality and protects water quality. *J. Soil Sci. Soc. Am.* 70:2114–2121.

Lalande, R.B. Gagnon, and R.R. Simard. 2003. Papermill biosolids and hog manure compost affect short-term biological activity and crop yield of a sandy soil. *J. Can. Soil Sci.* 83:353–362.

Li, Y., E. Hanlon, G. O'Connor, J. Chen, and M. Silveira. 2010. Land application of compost and other wastes (by-products) in Florida: regulations, characteristics, benefits, and concerns. *HortTechnology* 20:41–51.

Loper, S., A.L. Shober, C. Wiese, G.C. Denny, and C.D. Stanley. 2010. Organic soil amendment and tillage affect soil quality and plant performance in simulated residential landscapes. *HortScience* 45:1522–1528.

Madejón, E., P. Burgos, R. López, and F. Cabrera. 2001. Soil enzymatic response to addition of heavy metals with organic residues. *J. Biol. Fertility Soils.* 34:144–150.

McConnell, D.B., A. Shiralipour, and W.H. Smith. 1993. Compost application improves soil properties. *BioCycle* 34(4):61–63.

McSorley, R., P.A. Stansly, J.W. Noling, T.A. Obreza, and J. Conner. 1997. Impact of organic soil amendments and fumigation on plant-parasitic nematodes in a southwest Florida vegetable field. *Nematropica* 12:181–189.

Montemurro, F., G. Convertini, D. Ferri, and M. Maiorana. 2005. MSW compost application on Tomato Crops in mediterranean conditions: effects on agronomic performance and nitrogen utilization. *J. Compost Sci. & Utilization* 13:234–242.

Navas, A., F. Bermudez, and J. Machin. 1998. Influence of sewage sludge application on physical and chemical properties of Gypsisols. *J. Geoderma* 87:123–135.

Ozores-Hampton, M. 2006. Soil and nutrient management: compost and manure, Chapter 3, pp. 36–40. In *Grower's IPM Guide for Florida Tomato and Pepper Production*, Univ. of Fla., Gainesville, 210 pp.

Ozores-Hampton, M. 2012. Developing a vegetable fertility program using organic amendments and inorganic fertilizers. *HortTechnology* 22:743–750.

Ozores-Hampton, M. 2017a. Guidelines for assessing compost quality for safe and effective utilization in vegetable production. *HortTechnology* 27:150.

Ozores-Hampton, M. 2017b. Impact of soil health and organic nutrient management on vegetable yield and quality. *HortTechnology* 27:162–165.

Ozores-Hampton, M., P. Roberts, and P.A. Stansly. 2012. Organic pepper production, pp. 165–175. In V. Russo (ed.), *Peppers: Botany, Production and Uses*, CABI, Oxfordshire, UK.

Ozores-Hampton, M.P., P.A. Stansly, and T.P. Salame. 2011. Soil chemical, biological and physical properties of a sandy soil subjected to long-term organic amendments. *J. Sustainable Agr.* 353:243–259.

Ozores-Hampton, M., P.A. Stansly, R. McSorley, and T.A. Obreza. 2005. Effects of long-term organic amendments and soil solarization on pepper and watermelon growth, yield, and soil fertility. *HortScience* 40:80–84.

Ozores-Hampton, M.P., T.A. Obreza, and G. Hochmuth. 1998a. Composted municipal solid waste use on Florida vegetable crops. *HortTechnology* 8:10–17.

Ozores-Hampton, M., T.A. Obreza, G. Hochmuth. 1998b. Using composted wastes on Florida vegetables crops. *HortTechnology* 8:130–137.

Ozores-Hampton, M., and D.R.A. Peach. 2002. Biosolids in vegetable production systems. *HortTechnology* 12:356–340.

Ozores-Hampton, M., B. Schaffer, and H.H. Bryan. 1994b. Nutrient concentrations, growth, and yield of tomato and squash in municipal solid-waste-amended soil. *HortScience* 29:785–788.

Ozores-Hampton, M., P.A. Stansly, and T.A. Obreza. 2000. Biosolids and soil solarization effects on bell pepper (*Capsicum annuum*) production and soil fertility in a sustainable production systems. *HortScience* 35:443.

Pane, C., G. Celano, A. Piccolo, D. Villecco, R. Spaccini, A. Palese, and M. Zaccardell. 2015. Effects of on-farm composted tomato residues on soil biological activity and yields in a tomato cropping system. *J. Chem. Biol. Technol. Agri.* 2:1–13.

Perucci. 1992. Enzyme activity and microbial biomass in a field soil amended with municipal refuse. *Biol. Fertil. Soils* 14(1):54–60.

Press, C.M., W.F. Mahaffee, J.H. Edwards, and J.W. Kloepper. 1996. Organic by-products effects on soil chemical properties and microbial communities. *Compost Sci. Util.* 4(2):70–80.

Preusch, P.L., P.R. Adler, L.J. Sikora and T.J. Tworkoski. 2002. Nitrogen and phosphorus availability in composted and un-composted poultry litter. *J. Environ. Qual.* 31:2051–2057.

Price, G.W. and R.P. Voroney. 2007. Papermill biosolids effect on soil physical and chemical properties. *J. Environ. Qual.* 36:1704–1714.

Roe, N.E., and C.C. Gerald. 2010. Effects of dairy lot scrapings and composted dairy manure on growth, yield, and profit potential of double-cropped vegetables. *J. Compost Sci. Util.* 8:320–237.

Ros, M., S. Klammer, B. Knapp, K. Aichberger, and H. Insam. 2006. Long-term effects of compost amendment of soil on functional and structural diversity and microbial activity. *J. Soil Use Mgt.* 22: 209–218.

Salazar, I., D. Millar, V. Lara, M. Nuñez, M. Parada, M. Alvear, and J. Baraona. 2012. Effects of the application of biosolids on some chemical, biological and physical properties in an Andisol from southern Chile. *J. Soil Sci. Plant Nutr.* 12:441–450.

Serra-Wiltling, C., S. Houot, and E. Barriuso. 1996. Modification of soil water retention and biological properties by municipal solid waste. *Compost Sci. Util.* 4(1):44–52.

Shiralipour, A. 1998. The effects of compost on soil, pp. 27–31. In Florida Department Environmental Protection (ed.), *Compost Use in Florida.* 27 Nov. 2020 https://floridadep.gov/OrganicsRecycling

Sikora, L.J. and N.K. Enkiri. 2003. Availability of poultry litter compost P to fescue compared with triple super phosphate. *Soil Sci.* 168:192–199.

Stoffella, P.J., Z.L. He, S.B. Wilson, M. Ozores-Hampton, and N.E. Roe. 2014. Utilization of composted organic wastes in vegetable production systems. Food & Fertilizer Technology Center, Technical Bulletins. Taipei, Taiwan ROC. 3 May 2020. http://www.agnet.org/htmlarea_file/library/20110808105418/tb147.pdf.

Tester, C.F. 1990. Organic amendment effects on physical and chemical properties of a sandy soil. *Soil Sci. Soc. Am. J.* 54:827–831.

Tyler, R. 2001. Compost filter berms and blankets take on the silt fence. *BioCycle* 42(1):41–46.

U.S. Department of Agriculture (USDA). 2008. Soil quality indicators. Natural Resources and Conservation Services. 16 May 2020 https://www.nrcs.usda .gov/wps/portal/nrcs/detail/soils/health/assessment/?cid=stelprdb1237387

U. S. Environmental Protection Agency (USEPA), 1994. A plain English guide to the EPA part 503 biosolids rule. EPA832-R-93-003. Sept. Washington, DC.

U. S. Environmental Protection Agency (USEPA), 1995. A guide to the biosolids risk assessments for the EPA part 503 rule. EPA832-B-93-005. Sept. Washington, DC.

U. S. Environmental Protection Agency (USEPA), 1999. Biosolids generation, use, and disposal in the United States. EPA503-R-99-009. Sept. Washington, DC.

Warman, P.R. 1998. Results of long-term vegetable crop production trials: conventional vs compost-amended soils. *ISHS. Acta Hort.* 469:333–340.

Warman, P.R., and K.A. Havard. 1996. Yield, vitamin and mineral content of four vegetables grown with either composted manure or conventional fertilizer. *J. Veg. Crop Prod.* 2:13–25.

chapter 3

Fertility program using compost in fruit crops, vegetable and field ornamental production

Monica Ozores-Hampton

Contents

Fruit crop, vegetable and field ornamental production systems in the United States include open bed or plasticulture (Ozores-Hampton et al., 2015). These production systems have been effective as a standard commercial production, resulting in positive economic return. Most fruit crops or field ornamentals are trees, bushes or vines which will produce for years, and compost application to these crops is different than annual vegetable crops. For vegetables, large-seeded crops are typically seeded directly in open beds, whereas transplants (grown in a greenhouse in Styrofoam trays) are usually used for small-seeded crops, may be established directly on raised beds. Raised beds may be covered with a polyethylene mulch or remain uncovered (Ozores-Hampton et al., 2015). Conventional fruit crop, vegetable and ornamental growers, however, rarely add compost since the use of concentrated synthetic fertilizers are relatively inexpensive compared to the value of the crop and are readily available, resulting in high yields with maximum short-term profits (Kelly, 1990). The most common organic amendments that conventional fruit crop, vegetable and field ornamental growers can use are cover crops, animal raw manures and compost (Ozores-Hampton et al., 2015).

Incorporating cover crops into fruit crops, vegetable and field ornamental production may enhance the sustainability of the system by recycling unused nutrients from previous crops, improve soil structure, increase soil organic matter (SOM) and fertility, retain moisture, prevent leaching of nutrients, decrease soil density, suppress weeds, increase the population of beneficial insects, control erosion, manage plant parasitic nematodes, increase soil biological activity and increase yields (Abdul-Baki et al., 1997a,b; McSorley, 1998; Sainju and Singh, 1997; Stivers-Young, 1998; Sullivan, 2003; Treadwell et al., 2008a). Some benefits may occur during the cover crop life cycle, while other benefits may take effect after the cover crop is incorporated (Treadwell et al., 2008b). However, the disadvantages of growing cover crops within the fruit, vegetable and field crop production systems include additional production cost, delayed planting in annual systems, increased pest pressure, immobilization of fertilizer nitrogen (N) and the difficulty to control or remove such as vegetable crop (Treadwell et al., 2008c).

Raw manures can also supply plant macro and micronutrients and SOM. Increasing SOM improves soil structure and tilth, increases the water holding capacity, improves drainage, provides a source of slow-release nutrients, reduces wind and water erosion and promotes the growth of earthworms and other beneficial soil organisms (Rosen and Bierman, 2005). However, in areas of intense animal production, overfertilization with animal manure often occurs (Paik et al., 1996). The result is often manifested by nutrients entering adjacent water bodies. To obtain the maximum economic value of plant nutrients in animal manure and to protect water supplies from excessive nutrient runoff or leaching, animal manure should be applied to match the most environmentally limiting nutrient needs of a crop. In some states, the application of higher animal manure rates than the most limiting environmentally sensitive nutrients that are required by the crop (N or phosphorous (P) is illegal. The remaining nutrient amount, if any, must be supplied using synthetic fertilizers.

Therefore, compost application has advantages compared to cover crops and animal manures. Compost can be defined as "the product of a managed process through which microorganisms break down plant and animal materials into more available forms suitable for application to the soil" (Florida Department of Environmental Protection (FDEP), 1989). The technological and scientific advances in compost production, utilization, microbiology, and engineering (that has occurred during the past two decades, and implications that compost is environmental "friendly" and sustainable) have been reasons for the tremendous increase in worldwide compost usage (DeBertodi et al., 1987; Epstein, 1997; Ozores-Hampton et al., 1998, 2005a,b, 2011; Insam et al., 2010; Haug, 1993).

In areas of high population, there are a variety of nonhazardous wastes suitable for composting and land application that can provide an economically sound and environmentally acceptable option for utilization, but the majority of these wastes are currently landfilled or burned (Ozores-Hampton et al., 1998). Organic feedstocks composed of wastes produced by urban populations include: municipal solid waste (MSW); yard trimming (YT); food wastes from restaurants, grocery stores and institutions; wood wastes from construction and/or demolition; wastewater (from water treatment plants); and biosolids (sewage sludge). Agriculture produces other organic wastes that can be composted: poultry, dairy, horse, feedlot and swine manures; wastes from food processing plants; spoiled feeds; and harvest wastes (Ozores-Hampton et al., 1998, 2005a,b, 2017). The use of compost may improve soil quality and enhance the utilization of fertilizer, thus improving the performance of fruit, vegetable and ornamental crops (Ozores-Hampton et al., 1998, 2011; Ozores-Hampton and Peach, 2002). Also, compost application may control weeds (Ozores-Hampton et al., 2001a,b), suppress plant diseases (Hoitink and Fachy, 1986; Hoitink et al., 2001), increase SOM, decrease erosion by water and wind (Tyler, 2001) and reduce nutrient leaching (Jaber et al., 2005; Yang et al., 2007). The increase in SOM improves the physical properties of the soil by decreasing bulk density and increasing available water holding capacity (AWHC); chemical properties by increasing cation exchange capacity (CEC), pH and macro and micronutrient supplies (Ozores-Hampton et al., 2011; Sikora and Szmidt, 2001); and biological properties by increasing soil microbial activity properties (Ozores-Hampton et al., 2011).

In fruit crops and field ornaments, compost may be tilled into the soil as crops are initially established. Compost needs to be thoroughly incorporated so that changes in soil texture do not interfere with the root growth and the water movement. In clay soil, high compost applications can impede drainage and result in high moisture content in the root zone. After the initial plant establishment, compost can be added annually as a mulch. Compost mulches are particularly effective in dry areas, where there is no irrigation available. The mulch should be at least 2 to 4 inches thick and coarse enough to suppress weed growth. Mulch should be kept several inches from tree trunks to avoid increasing crown fungal diseases. In vegetable production, compost can be applied at the beginning of the growing season directly to planting beds and then tilled into the soil.

There are no U.S. government restrictions on how and when compost can be used in fruit crop, vegetable and field ornamentals production except for compost derived from sewage sludge or biosolids (Ozores-Hampton and Peach, 2002; U.S. Environmental Protection Agency (USEPA), 1994, 1995, 1999). However, to eliminate or reduce human and plant pathogen, nematodes and weeds, the temperature during the compost process must remain between 131° and 170° for 3 days in an in-vessel or static aerated pile, or 15 days in windrows, which must be turned at least five times during this period (USEPA, 1994, 1995, 1999).

3.1   Compost nitrogen availability in crop production

Compost N mineralization rates or N availability vary based on compost feedstocks, soil characteristics and environmental conditions. Compost N mineralization is the process by which microorganisms convert organic N into plant-available inorganic (NH_4-N and NO_3-N) forms (Diacono and Montemurro, 2010). Compost N mineralization occurs in phases, the first of which is usually rapid. In phase one, a significant release of inorganic N (in addition to that already present in the waste as NH_4-N and NO_3-N) can occur within a few weeks of soil application. The initial release is followed by slower decomposition over time. Composted materials may take several years to decompose in the field. "Decay series" have been used to describe N mineralization or bioavailability during a period of several years after application. A decay series for a specific compost material can be used to predict the amount of N that will be available to a crop each year. Therefore, the rate of N release is especially important since N nutrient moves readily through sandy soils. Evaluations of N mineralization in situ can be used to improve N use efficiency. However, quantitative measurement of N mineralization in situ is difficult due to the complex and dynamic N transformations in the soil environment (Preusch et al., 2002). More than 90% of the total N in compost will be in an organic form and only 10% will be in the inorganic forms of nitrate (NO_3-N) or ammonium (HN_4-N) (Hartz et al., 2000). Therefore, the application time may not be as critical when compared to raw animal manures as quality compost has many slow-release nutrient benefits. The composting process converts raw organic materials such as raw manure high in NH_4 and NO_3, which is susceptible to runoff or leaching, to a humus-stable form, minimizing the environmental impact on air and groundwater contamination.

Nitrogen immobilization occurs in composts with the initial carborn:nitrogen (C:N) ratio greater than 20:1, and mineralization occurs when composts had a C:N ratio lower than 20:1. However, the C:N ratio as a predictor of N mineralization is not exact, since it may depend on the type of C (Prasad, 2009a; Rosen and Bierman, 2005; Wallace, 2006). Mineralization N rates guidelines developed by Wallace (2006) indicated that the availability of N will be 0%–20% or even negative in the first year and 0%–8% in the following years. The decrease in C:N ratio throughout the composting process will result in the net N mineralization rate (Van Kessel et al., 2000). Nitrogen mineralization and immobilization process are associated to the C-cycle, as the energy supplies for microorganisms develop from C compounds (Deenik, 2006). The pH of the composted solution plays a factor by regulating the stability between NH_4 and NH_3 (Nahm, 2005). With a pH value below 7.0, most of the NH_3-N is tied up as NH_4-N and will not be susceptible to volatilization. Electrical conductivity (EC) and pH can affect the growth of bacteria and fungi while the OM content and C:N ratio influence the growth of actinomycetes (Gebeyehu and Kibret, 2013; Rebollido et al., 2008). If animal manure is composted, the organic N content will be lower than in raw animal manures. Raw animal manure contains more NH_4-N content than compost, which increases the risk of volatilization to ammonia (NH_3) gas. Therefore, raw animal manure should be field incorporated within 12 hours of application to decrease NH_3-N losses (Rosen and Bierman, 2005). It is important to know that the N mineralization (decomposition or microbial breakdown) rate of the compost before determining the application rate to crops (Table 3.1). Nitrogen mineralization of stable and mature compost can range from 1% to 20% depending on feedstocks, soil and environmental conditions (temperature). Soil temperature can influence N mineralization, as soil temperature increases N mineralization and decreases as soil temperatures decreases. Compost N content are low and not considered to be a major source of N with low environmental impact. However, feedstocks having high N content together with high compost rate application may negatively impact the environment.

3.2 Compost phosphorous and potassium availability in crop production

There are limited phosphorus (P) studies evaluating the dynamic of composted P and potassium (K) in a fertility program in crop production. Compost P and K availability vary based on compost feedstocks, soil characteristics and environmental conditions. In acidic soils with a pH of less than 6.0, P availability are predominantly controlled by aluminum (Al) and iron (Fe), (Gagnon et al., 2012) and as soil pH increases, the potential for P precipitation increases (Mkhabelaa and Warman, 2005). Although compost improves P availability via organic, humic and fulvic acids (Gourango, 2007; Minggang et al., 1997), the organic acid compounds released during OM decomposition are responsible for increasing negative charges (raising pH) on Al and Fe oxide surfaces which reduces the number of available binding sites for other metallic ions such as plant nutrients through competition and solubility (Admas et al., 2014). Soil type has also been shown to influence inorganic P sorption capacity (Preusch et al., 2002; Chen et al., 2001, 1994). The fate of composts applied to soils, which results in initial immobilization or mineralization of soil P, will depend on the C:P ratio of the stable organic feedstock applied

Table 3.1 The effect of compost feedstock source, rate and temperature using laboratory incubation method on nitrogen (N) mineralization

Compost type[x]	Study temperature (°F)	Compost rate (tons/acre)	Incubation time (days)[z]	Initial total N (%)	NO$_3$ + NH$_4$ (ppm)	N Mineralization[y] (%)	Reference
PM/wood chips	77	-	120	5.0	638	1.4	Preusch et al., 2002
Pelletized PM/ST	59	10.5	150	4.0	1,456	3.6	Flavel and Murphy, 2006
Straw/SWM/SD	59	10.5	150	1.4	381	2.7	Flavel and Murphy, 2006
YW/PM	59	10.5	150	1.4	279	2.0	Flavel and Murphy, 2006
Broiler PM/hay	77		56	1.3	730	5.6	Tyson and Cabrera, 1993
PM/leaves	77		56	1.0	550	5.5	Tyson and Cabrera, 1993
BS	35		180	2.2	4,335	19.7	Lindemann and Cardenas, 1984
BS	77	61	112	2.4	2,958	12.3	Huang and Chen, 2009
DM	86	36	180	5.0	4,889	9.8	Hadas et al., 1986
BS/cotton waste	82	19	70	3.8	1,261	3.3	Bernal et al., 1998
MSW/YW	59		149	1.6	1,365	8.5	Luxhøi et al., 2007
SM/ST	77		130	1.7	2,910	7.7	N'Dayegamiye et al., 1997
DM/ST	68	24	84	2.0	413	2.1	Shi et al., 1999
MSW	82	20	120	1.2	820	6.8	Madejón et al., 2001
MSW	82		270	1.9	3,400	8.0	Beloso et al., 1993
DM/PM/ST	86		180	0.2	1,520	7.6	Hadas and Portnoy, 1994
BS	73	12	112	4.3	2,291	5.3	Parker and Sommers, 1983

[z] Data were considered only for ≤180 incubation days.

[y] N mineralization (%) = [(nitrate (NO$_3$) + ammonium (NH$_4$) ppm)/(initial total N (%) × 10^4 ppm)] × 100.

[x] BS = biosolids; DM = dairy manure; MSW = municipal solid waste; N = nitrogen; PM = poultry manure; SD = sawdust; SM = sheep manure; ST = straw; SWM = swine manure; YW = yard waste.

(Prasad, 2013). Hence, P immobilization could be prevalent in C:P ratio of more than 200 (Hannapel et al., 1964). However, studies showed that slightly lower P availability from composted manure than non-composted manure indicated a chemical reaction of P during composting, which may cause P to become less plant available (Eghball et al., 2002). Proposed P availability from various compost made from different feedstock relative to superphosphate are available for spent mushroom compost (100%), animal manures (90%), sewage sludge (85%), source separated food waste (75%) and yard waste (60%) (Prasad, 2009b).

P and K will not react with N when compost is added to soil. P and K in compost, once is incorporated in the soil, is readily available for plant uptake, because the OM content in the compost blocks sites where P will be adsorbed. In addition of the compost to soil, the biological activity can cause the release of soil-bound P, resulting in a net P uptake by the crop. There were no differences between compost and commercial P in studies using biosolids compost or animal manure compost (Preusch et al., 2002; Sikora and Enkiri, 2003). Hence a compost end user should be cautious when using compost as N fertilizer because only a portion of the N (5%–30%) will behave as a commercial fertilizer the first year, but all the P and K in the compost will react as a commercial fertilizer. Therefore, compost application on P-sensitive land should be calculated based on P crop rather than N crop requirements (Preusch et al., 2002; Sikora and Enkiri, 2003). However, generally for compost, 70–80% of the P and 80–90% of the K will be available from animal manure during the first year after application (Rosen and Bierman, 2005).

3.3 Compost micronutrient availability in crop production

Compost can provide the following essential micronutrients in an available form: boron (B), copper (Cu), iron (Fe), manganese (Mn), molybdenum (Mo) and zinc (Zn) (Diacono and Montemurro, 2010). While micronutrient crop requirements are in small quantities when compared to macronutrients, they are essential to plant growth and development, and many soils are low on these elements. The micronutrient content of compost varies widely depending on the feedstocks, soil and environmental conditions.

3.4 Effects of compost heavy metal content in crop production

Heavy metals are toxic and persistent pollutants that may be present in feedstock materials for compost and are used as soil amendments (Ozores-Hampton et al., 2005b). Toxic metals may accumulate in the soil (Sterrett et al., 1982; Yuran and Harrison, 1986) or are uptaken by the plant and accumulate in the edible plant parts, where they pose a potential threat to consumers (Shiralipour et al., 1992). Metals that pose the greatest threat to human health are Cd, Cu, Pb, Ni and Zn (Chaney, 1993). Presently, biosolids and biosolids mixed with either YTW or MSW are regulated at the federal level under the Clean Water Act Section 503 (USEPA, 1994, 1995). The Clean Water Act Section 503 classified the quality of two biosolids based on the nine regulated pollutant elements concentration limit; the pollutant ceiling concentration and pollutant concentration; and two loading rates–based limit, cumulative pollutant loading rates (CPLR) and annual pollutant loading rates (APLR) (USEPA 1994, 1995). Eighteen states have regulations in place that are more restrictive than Section 503 (Goldstein, 2000). Evaluation of potential food-chain transfer of Cd, Cu, Pb, Ni and Zn in compost shows that consumption for 70 years of 60% of garden food, produced on pH 5.5 soils and amended with 446 tons/acre of compost, would be safe (Chaney, 1994). Biosolids or compost made from waste materials that do not meet EPA 503 standards for pollutant concentration should not be applied to horticultural land (Ozores-Hampton and Peach, 2002). Therefore, repeated, long-term applications of compost made from waste materials with pollutant content below maximum acceptable levels under state and federal regulations should be suitable for vegetable production (Ozores-Hampton et al., 1994a,b; Ozores-Hampton et al., 1997).

3.5 Accumulation of soluble salts and compost application in crop production

The electrical conductivity of some compost can raise a concern with repeated applications over a long period of time. Furthermore, high compost rates can proportionally increase EC soil concentrations. High EC concentration can inhibit germination of direct seeded crops and growth and development in sensitive species. Normally, EC will decline over time due to crop removal and leaching by rainfall. However, in desert areas where rainfall is limited, EC accumulation due to high compost in EC and/or with high application rates needs to be considered annually. Some of the EC in animal manures is from the nutrient salts, such as K, Ca and Mg. However, compost high in sodium is not desirable in crop production.

3.6 Using compost in a fertility program in crop production

The fertility program for conventional and organic fruit orchards, vegetable and ornamental production can be divided into two major parts: an organic-amendment-based program and supplemental fertility program consisting of inorganic or organic fertilizer such as ammonium nitrate, urea and potassium sulfate, plus micronutrients to supply the plant nutrients requirements.

Those using composts must practice sound soil fertility management to prevent nutrient imbalances and associated health risks, as well as surface-water and groundwater contamination. Matching compost supplied nutrients to vegetable nutrient requirements should be the goal of a conventional vegetable fertility program (Ozores-Hampton et al., 2011). Overfertilization will be inefficient and expensive, which may contribute to nutrient runoff, groundwater pollution, soil toxicity, pest and disease susceptibility, excessive production of foliage and reduced vegetable quality and yields. Similarly, under-fertilization can reduce vegetable yield and/or quality.

Table 3.1 provides an analysis of compost suitable for fruit orchards, vegetable and ornamental production. Since actual nutrient content varies considerably between compost sources, a representative product sample should be sent to a laboratory for analysis of moisture and nutrient content such as total N, phosphate (P_2O_5), potash (K_2O), calcium (Ca), magnesium (Mg) and micronutrients. Additionally for compost, nitrate-N (NO_3-N) and ammonium-N (NH_4-N) is recommended. Accurate compost analysis requires that a representative sample be submitted, so several subsamples should be collected and combined for analysis.

For successful integration of compost into conventional vegetable fertility programs, we recommend the construction of an N-P-K crop mass-balance where the fertility inputs and net release of N will be quantified and vegetable crop N-P_2O_5-K_2O requirement will be taken into consideration. Calculating N availability from compost can be complex since N must be transformed by soil microorganisms before it can be utilized by the vegetable crop as NO_3-N. An example of a tomato fertility program for Florida is provided as a guide in Table 3.1.

The first step in building the tomato fertility program is to determine the tomato crop nutrient requirements by taking a soil sample for analysis of N-P-K and micronutrients. These results can be compared to the local crop recommendations for N-P_2O_5-K_2O. This information can be found in a local state or regional "vegetable production handbook" or by contacting a local extension faculty. Then, the identification of locally available compost. Once the compost was located and identified, determine the nutrient content and N, P and K availability from laboratory analysis or from other sources such as given in Table 3.2. The microbial activity involved in the N cycles, which is accelerated by high temperatures and slowed down with low temperatures, needs to be considered and N release rate adjusted as need. Then, the next step will

Table 3.2 Nitrogen (N), phosphorus (P) and potassium (K) concentrations and N
mineralization rates of compost for vegetable crop production

Organic feedstocks[z]	N	P	K	Rate of N release (%/year)
	------------ (%) -----------			
Biosolids	3–6	2–3	0.10–0.15	3.0–20
Brewery waste solids	1.3–1.8	0.02	0.13–0.18	5.0–10
Dairy manure	1.2–1.5	0.3	0.9	6.0–15
Feedlot manure	1.9–2.2	0.3–1.2	0.6–3.2	3.0–15
Fruit and vegetable waste	1.39	0.26	1.19	10
Gin trash	1.2–3.8	0.2	1.2	10
Horse manure	0.5	0.2	0.4	10
Food waste	1.1–1.8	0.03–0.09	0.35–0.45	2.0–12
Municipal solid waste	2.3	1.11	0.64	3.0–10
Mushroom substrate	2.5	1.3	0.9	10
Olive mill waste	3.5	0.17	2.3	20
Poultry manure	1.3–5	3.0	2.0	20
Yard waste	1.0–1.2	0.2–0.3	0.2–1.4	2.0–10

[z] Altieri and Esposito, 2010; Ahmad et al., 2008; Aram and Rangarajan, 2005; Bellows, 2003; Chellemi and Lazarovits, 2002; Diver et al., 1999; Drinkwater, 2007; Gaskell, 2009; Gaskell and Smith 2006, 2007; Gaskell and Klauer, 2004; Hartz and Johnstone, 2006; Hartz et al., 2000; Kuepper and Everett 2004; Prasad, 2009a,b; Pressman, 2009; Sooby et al., 2007; Rosen and Bierman, 2005; VanTine and Verlinden, 2003; Zhang and Li, 2003.

be the calculation of compost application rates to supply recommended amounts of N, P_2O_5 and K_2O to the tomato crop so that yield estimates are realized (Tables 3.2 and 3.3). To calculate the correct application rate of compost, multiply by availability factors (70%–80% for P and 80%–90% for K) to obtain the amount of P and K that will be available to vegetables from the application of compost. Then, multiply the total P by 2.29 and K by 1.2 to obtain P_2O_5 and K_2O (Table 3.3). The advantage of using compost rather than raw manure will be that, although P can be over-applied with compost, the improvement in soil structure with the compost OM application will increase water infiltration and reduce run-off, thereby decreasing the total P transported over the land surface to potentially pollute surface water (Spargo et al., 2006). Finally, determine whether application of inorganic commercial fertilizer is needed. Once a fertility program is established, a unit cost per nutrient can be calculated. The cost per unit of nutrients can be calculated by multiplying the nutrient fertilizer unit cost by the organic amendment available nutrient content and then by selecting the most cost-effective one to be applied to the tomato crop.

Table 3.3 Florida nutrient mass budget for conventional tomato production[z]

Material inputs	Application rate (lb/acre DW)[y]	N rate (lb/acre)	N mineralization rate (%)	Total (lb/acre NO₃)	Total (lb/acre P₂O₅)[x]	Total (lb/acre K₂O)[x]
Yard waste compost at 4.0 ton/acre (40% moisture and 1% N, 0.2% P and 0.8% K with 70% P and 80% K availability)	6,725	67.2	10	6.7	21.5	43.0
Inorganic fertilizer application						
Ammonium nitrate[w]	568.5			193.3	-	-
Triple phosphate	170.7			-	78.5	-
Potassium sulfate	105.6			-	-	57.0
Total	-			200	100	100

[z] Tomato (*Solanum lycopersicum*) nutrient requirements based on 200 lb/acre of nitrogen (N), 100 lb/acre phosphoric acid (P₂O₅) and potassium oxide (K₂O) with a medium soil test levels of P and K, respectively (Olson et al., 2010).

[y] DW = dry weight; NO₃ = nitrate.

[x] P × 2.2910 = P₂O₅, K × 1.2047 = K₂O.

[w] Ammonium nitrate (34% N); triple phosphate (46% P₂O₅); potassium sulfate (54% K₂O).

Environmental monitoring has shown elevated NO_3-N or P concentration to be widespread in both surface and groundwater, often occurring in regions with concentrated horticultural production systems. Compost applications have many positive effects on the soil and agricultural production system. Routine applications often result in increased soil bulk density, AWHC, improved soil structure, increased soil carbon content, additional macro and micronutrients, buffered pH, reduced soluble salts, increased CEC and increased biological activity and diversity (microbial biomass). Also, compost use can improve water quality in high production areas by improving the soil and decreasing the use of highly soluble synthetic inorganic fertilizers. However, a fertility program using compost requires the understanding of the contribution of nutrients such as N, over time, P, K and other micronutrients that are present in most compost sources. Higher compost application rates for soil conditioning may produce excessive nutrient buildup in the soil and nutrient loss to the environment if the compost program is not carefully planned. In dry climates with low opportunity for nutrient loss due to leaching, high compost application rates can produce excessive salt and P and K soil accumulation that can interfere with plant growth, nutrient uptake or cause a deficiency of other nutrients. The first step in building a conventional fertility program will be to take a soil sample and send it to a soil laboratory for a nutrient analysis. These results should be compared to the local crop recommendations. Second, select the compost based on local availability. Then, determine the nutrients available from the compost and use synthetic inorganic fertilizer sources to satisfy the crop nutrient requirements not supplied from the compost source.

World populations are exploding, and the ever increasing need for more food supplies is driving more land area into intensive agricultural production. In parallel, global waste production is increasing, and therefore connecting waste streams to horticulture production throughout compost utilization can increase crop production without impacting the environment.

References

Abdul-Baki, A.A., J.R. Teasdale, and R. Korcak. 1997a. Nitrogen requirement of fresh-market tomatoes on hairy vetch and black polyethylene mulch. *HortScience* 32:217–221.

Abdul-Baki, A.A., R.D. Morse, T.E. Devine, and J.R. Teasdale. 1997b. Broccoli production in forage soybean and foxtail millet cover crop mulches. *HortScience* 32:836–839.

Admas, H., G. Heluf, B. Bobie, and A. Enyew. 2014. Effects of compost and mineral sulfur fertilizers on phosphorus desorption at Wujiraba watershed, Northwestern highlands of Ethiopia. *J. Global Sci. Frontier Res.* 14:69–78.

Ahmad, R., M. Naveed, M. Aslam, Z.A. Zahir, M. Arshad, and G. Jilani. 2008. Economizing the use of nitrogen fertilizer in wheat production through enriched compost. *Renewable Agr. Food Systems* 23:243–249.

Altieri R., and A. Esposito. 2010. Evaluation of the fertilizing effect of olive mill waste compost in short-term crops. *Intl. Biodeterior. Biodegrad.* 64:124–128.

Aram, K., and A. Rangarajan. 2005. Compost for nitrogen fertility management of bell pepper in a drip-irrigated plasticulture system. *HortScience* 40:577–581.

Beloso, M.C., M.C. Villar, A. Cabaneiro, M. Carballas, S.J. Gonzfilez-Prieto, and T. Carballas. 1993. Carbon and nitrogen mineralization in an acid soil fertilized with composted urban refuses. *J. Bioresource Technol.* 45:123–129.

Bellows, B. 2003. Protecting water quality on organic farms, pp. 1–36. In J. Ridle and J. Ford (eds.), *Organic Crops Workbook: A Guide to Sustainable and Allowed Practices.* Natl. Ctr. Appropriate Technol., Fayetteville, AR.

Bernal, M.P., A.F. Navarro, M.A. Sanchez-Monedero, A. Roig, and J. Cegarra. 1998. Influence of sewage sludge compost stability and maturity on carbon and nitrogen mineralization in soil. *J. Soil Biol. Biochem.* 30:305–313.

Chaney, R.L. 1993. Risks associated with the use of sewage sludge in agriculture. *Proceeding of Federal Australian Water and Wastewater Association,* Gold Coast, Queensland, April 18–23. Vol. 1. Australian Wastewater Association Branch, West End, Queensland, Australia.

Chaney, R.L. 1994. Trace metal movement: soil-plant systems and bioavailability of biosolids-applied metals, pp. 27–54. In C.E. Clapp, W.E. Larson, and R.H. Dowdy (eds.), *Sewage Sludge: Land Utilization and the Environment,* American Society of Agronomy.

Chellemi, D.O. and G. Lazarovits. 2002. Effect of organic fertilizer applications on growth, yield and pests of vegetables crops. *Proc. Florida State Hort. Soc.* 115:315–321.

Chen, J.H., Y.E. Weng, and Y.P. Wang. 1994. Effects of organic fertilizers addition on P sorption characteristics of soils. *J. Chinese Agr. Chem. Soc.* 3:332–346.

Chen, J.H., J.T. Wu, and W. Tin. 2001. Effects of compost on the availability of nitrogen and phosphorus in strongly acidic soils. Huang Department of Agricultural Chemistry, Taiwan Agricultural Research Institute, Wufeng, Taiwan ROC.

DeBertodi, M., M.P. Ferranti, P. L'Hermite, and F. Zucconi. 1987. *Compost: Production, Quality, and Use,* Elsevier Appl. Sci., London, U.K.

Deenik, J. 2006. Nitrogen mineralization potential in important agricultural soils of Hawaii. College of Tropical Agriculture and Human Resources (CTAHR). University of Hawaii. Mänoa, USA. 24 April 2020. http://www.ctahr.hawaii.edu/deenikj/Downloads/SCM-15.pdf.

Diacono, M., and F. Montemurro. 2010. Long-term effects of organic amendments on soil fertility: a review. *Agron. Sustainable Dev.* 30(2):401–422.

Diver, S., G. Kuepper, and H. Born. 1999. Organic tomato production. Appropriate technology transfer for rural areas. Natl. Sustainable Agr. Information Ctr.

Drinkwater, L. 2007. On-farm budgets in organic cropping systems: A tool for fertility management. Organic Farming Res. Foundation, Santa Cruz, CA.

Eghball, B., B.J., Wienhold, J.E. Gilley, and R.A. Eigenberg. 2002. Mineralization of manure nutrients. *J. Soil and Water Conserv.* 57:470–473.

Epstein, E. 1997. *The Science of Composting.* CRC Press, Boca Raton, FL.

Flavel, T.C., and D.V. Murphy. 2006. Carbon and nitrogen mineralization rates after application of organic amendments to soil. *J. Environ. Quality* 35:183–193.

Florida Department of Environmental Protection (FDEP). 1989. Criteria for the pro-
    duction and use of compost made from solid waste. Florida Administrative
    Code, Chapter 17-709. Tallahassee, FL.

Gagnon, B., I. Demers, N. Ziadi, M.H. Chantigny, L.E. Parent, T.A. Forge, F.J.
    Larney, and K.E. Buckley. 2012. Forms of phosphorus in composts and in
    compost-amended soils following incubation. *J. Can. Soil Sci.* 92:711–721.

Gaskell, M. 2009. In-season fertilization in organic systems. Univ. California
    Coop. Ext. 24 April 2020. http://cesantabarbara.ucdavis.edu/files/75469.pdf.

Gaskell, M., and H. Klauer. 2004. The effect of green manure, compost and feather
    meal on soil nitrogen dynamics and bell pepper yield. Organic Farming Res.
    Foundation, Santa Cruz, CA.

Gaskell, M., and R. Smith. 2006. Soil fertility management for organic crops. Univ.
    California Coop. Ext.

Gaskell, M., and R. Smith. 2007. Nitrogen sources for organic vegetable crops.
    Univ. California Coop. Ext.

Gebeyehu, R., and Kibret, M. 2013. Microbiological and physico-chemical analysis
    of compost and its effect on the yield of kale (Brassica oleracea) in Bahir Dar,
    Ethiopia. *J. Ethiopian Sci. Technol.* 6:93–102.

Goldstein, N. 2000. The state of biosolids in America. *BioCycle* 41(12):50–56.

Gourango, k. 2007. Phosphorus speciation in biosolids amended soils: correlating
    phosphorus desorption, sequential chemical extractions, and phosphorus-
    xanes spectroscopy, University of Saskatchewan Saskatoon, Saskatchewan,
    Canada.

Hadas, A., and R. Portnoy. 1994. Nitrogen and carbon mineralization rates of com-
    posted manures incubated in soil. *J. Environ. Quality* 23:1184–118.

Hadas, A., L. Kautsky, and R. Portnoy. 1986. Mineralization of composted manure
    and microbial dynamics in soil as affected by long-term nitrogen manage-
    ment. *J. Soil Biol. Biochem.* 28:733–738.

Hannapel, R.J., W.J. Fuller, and R.J. Fox. 1964. Phosphorus movement in calcareous
    soil. *J. Soil Sci.* 97:421–427.

Hartz, T.K., and P.R. Johnstone. 2006. Nitrogen availability from high-nitrogen-
    containing organic fertilizers. *HortTechnology* 16:39–42.

Hartz, T.K., J.P. Mitchell, and C. Giannini. 2000. Nitrogen and carbon mineraliza-
    tion dynamics of manures and composts. *HortScience* 35:209–212.

Haug, R.T. 1993. *The Practical Handbook of Compost Engineering*. Lewis Publishers,
    Boca Raton, FL.

Hoitink, H.A., and P.C. Fachy. 1986. Basis for the control of soil-borne plant patho-
    gens with composts. *Annu. Rev. Phytopathol.* 24:93–114.

Hoitink, H.A., M.S. Krause, and D.Y. Han. 2001. Spectrum and mechanisms of
    plant disease control with composts, pp. 263–274. In P.J. Stoffella and B.A.
    Kahn (eds.), *Compost Utilization in Horticultural Cropping Systems*, CRC Press,
    Boca Raton, FL.

Huang, C., and Z. Chen. 2009. Carbon and nitrogen mineralization of sewage
    sludge compost in soils with a different initial pH. *J. Soil Sci. Plant Nutr.* 55:
    715–724.

Insam, H., N. Riddech, and S. Klammer (eds.). 2010. *Microbiology of Composting*.
    Springer-Verlag, Berlin, Germany.

Jaber, F.H., S. Shukla, P.J. Stoffella, T.A. Obreza, and E.A. Hanlon. 2005. Impact
    of organic amendments on groundwater nitrogen concentrations for sandy
    and calcareous soils. *Compost Sci. Utilization* 13:194–202.

Kelly, W.C. 1990. Minimal use of synthetic fertilizers in vegetable production. *HortScience* 25:168–169.

Kuepper, G. and K. Everett. 2004. Potting mixes for certified organic production. Appropriate Technol. Transfer Rural Areas-Natl. Sustainable Agr. Info. Ctr.

Lindemann, W.C. and M. Cardenas. 1984. Nitrogen mineralization potential and nitrogen transformations of sludge-amended soil. *J. Soil Sci. Soc. Am.* 48:1072–1077.

Luxhøi, J., S. Bruun, L.S. Jensen, J. Magid, A. Jensen, and T. Larsen. 2007. Municipal solid waste Modelling C and N mineralization during decomposition of anaerobically digested and composted municipal solid waste. *J. Waste Mgt. Res.* 25:170–176.

Madejón, E., P. Burgos, R. López, and F. Cabrera. 2001. Soil enzymatic response to addition of heavy metals with organic residues. *J. Biol. Fertility Soils* 34:144–150.

Marr, C.W., R. Janke, and P. Conway. 1998. Cover crops for vegetable growers. Kansas State Univ. Coop. Ext. Serv. MF 2345.

McSorley, R. 1998. Alternative practices for managing plant-parasitic nematodes. *Amer. J. Alternative Agr.* 13:98–104.

Minggang L., M. Osaki, I.M. Rao, and T. Tadano. 1997. Secretion of phytase from the roots of several plant species under phosphorus-deficient conditions. *J. Plant Soil* 195:161–169.

Mkhabela, M.S., and P.R. Warman. 2005. The influence of municipal solid waste compost on yield, soil phosphorus availability and uptake by two vegetable crops grown in a Pugwash sandy loam soil in Nova Scotia. *J. Agri. Eco. Environ.* 106:57–67.

Nahm, K.H. 2005. Factors influencing nitrogen mineralization during poultry litter composting and calculations for available nitrogen. *J. World's Poultry Sci.* 61:238–255.

N'Dayegamiye, A., R. Royer, and P. Audesse. 1997. Nitrogen-mineralization and availability in manure composts from Québec biological farms. *J. Can. Soil Sci.* 77:345–350.

Olson, S.M., W.M. Stall, G.E. Vallad, S.E. Webb, S.A. Smith, E.H. Simonne, E. McAvoy, and B.M. Santos. 2010. Tomato production in Florida, pp. 295–316. In: S.M. Olson and B. Bielinski (eds.), *2010–2011 Vegetable Production Handbook for Florida.* Vance Publishing, Lenexa, KS.

Ozores-Hampton, M. 2006. Soil and nutrient management: compost and manure, pp. 36–40. In: J.L. Gillett, H.N. Petersen, N.C. Leppla, and D.D. Thomas. *Grower's IPM Guide for Florida Tomato and Pepper Production*, Univ. Florida, Gainesville, FL.

Ozores-Hampton, M. 2017. Past, present, and future of compost utilization in horticulture. *Acta Horticulturae. Acta Horticulturae.* International symposium on growing media, soilless cultivation, and compost utilization in horticulture. 27 Nov. 2020 https://www.actahort.org/books/1266/1266_43.htm. (In press).

Ozores-Hampton, M., F. Di Gioia, S. Shinjiro, E. Simonne, K. Morgan. 2015. Effects of nitrogen rates on nitrogen, phosphorous and potassium partitioning, accumulation and use efficiency in seepage-irrigated fresh market tomatoes. *HortScience* 50:1636–1643.

Ozores-Hampton, M.P., E.A. Hanlon, H.H. Bryan and B. Schaffer. 1997. Cadmium, copper, lead, nickel, and zinc concentrations in tomato and squash in compost-amended calcareous soil. *Compost Sci. Util.* 5(4):40–45.

Ozores-Hampton, M.P., T.A. Obreza, and G. Hochmuth. 1998. Composted municipal solid waste use on Florida vegetable crops. *HortTechnology* 8:10–17.

Ozores-Hampton, M.P., and D.R. Peach. 2002. Biosolids in vegetable production systems. *HortTechnology* 12:18–22

Ozores-Hampton, M.P., P.A. Stansly, and T.P. Salame. 2011. Soil chemical, biological and physical properties of a sandy soil subjected to long-term organic amendments. *J. Sustainable Agr.* 353:243–259.

Ozores-Hampton, M.P., P.A. Stansly, R. McSorley, and T.A. Obreza. 2005a. Effects of long-term organic amendments and soil solarization on pepper and watermelon growth, yield, and soil fertility. *HortScience* 40:80–84.

Ozores-Hampton, M., P.A. Stansly, and T.A. Obreza. 2005b. Heavy metal accumulation in a sandy soil and in pepper fruit following long-term application of organic amendments. *Compost Sci. Util.* 13:60–64.

Ozores-Hampton, M., T.A. Obreza, and P.J. Stoffella. 2001a. Weed control in vegetable crops with composted organic mulches, pp. 275–286. In P. J. Stoffella and B. A. Kahn (eds.), *Compost Utilization in Horticultural Cropping Systems,* CRC Press, Boca Raton, FL.

Ozores-Hampton, M.P., T.A. Obreza, P.J. Stoffella, and G. Fitzpatrick. 2001b. Immature compost suppresses weed growth under greenhouse conditions. *Compost Sci. Utilization* 10:105–113.

Ozores-Hampton, M.P., B. Schaffer and H.H. Bryan. 1994a. Mineral elements concentrations, growth, and yield of tomato and squash in calcareous soil amended with municipal solid waste compost. *HortScience* 29:785–788.

Ozores-Hampton, M., B. Schaffer, and H.H. Bryan. 1994b. Influence of municipal solid waste (MSW) compost on growth, yield and heavy metal content of tomato (Abstract). *HortScience* 29:451.

Paik, I.K., R. Blair, and J. Jacob. 1996. Strategies to reduce environmental pollution from animal manure -principles and nutritional management - A Review. *Asian-Austral. J. Animal Sci.* 9:615–635.

Parker, C.F. and L.E. Sommers. 1983. Mineralization of nitrogen in sewage sludge. *J. Environ. Qual.* 12:150–156.

Prasad, M. 2009a. A literature review on the availability of nitrogen from compost in relation to the nitrate regulations SI 378 of 2006. Environmental Protection Agency, Wexford, Ireland.

Prasad, M. 2009b. A literature review on the availability of phosphate from compost in relation to the nitrate regulations SI 378 of 2006. Environmental Protection Agency, Wexford, Ireland.

Prasad, M. 2013. A literature review on the availability of phosphorus from compost in relation to the nitrate regulations SI 378 of 2006. Environmental Protection Agency. Johnstown Castle, Co. Wexford, Ireland.

Pressman, A. 2009. Sources of organic fertilizers & amendments. Appropriate Technol. Transfer Rural Areas-Natl. Sustainable Agr. Info. Ctr. 24 April 2020. https://attra.ncat.org/attra-pub/org_fert/.

Preusch, P.L., P.R. Adler, L.J. Sikora, and T.J. Tworkoski. 2002. Nitrogen and phosphorus availability in composted and un-composted poultry litter. *J. Environ. Qual.* 31:2051–2057.

Rebollido, R.J. Martínez, Y. Aguilera, K. Melchor, I. Koerner, and R. Stegmann. 2008. Microbial populations during composting process of organic fraction of municipal solid waste. *J. Appl. Ecol. Environ. Res.* 6:61–67.

Rosen, C.J. and P.M. Bierman. 2005. Using manure and compost as nutrient sources for fruit and vegetable crops. Univ. Minnesota Ext. Serv. M1192.

Sainju, U.M., and B.P. Singh. 1997. Winter cover crops for sustainable agricultural systems: Influence on soil properties, water quality, and crop yields. *HortScience* 32:21–28.

Shi, W., J.M. Norton, B.E. Miller, and M.G. Pace. 1999. Effects of aeration and moisture during windrow composting on the nitrogen fertilizer value of dairy waste composts. *Appl. Soil Ecol.* 11(3):17–28.

Shiralipour, A., D.B. McConnell, and W.H. Smith. 1992. Use and benefits of municipal compost: a review and assessment. *Biomass Bioener.* 3:267–279.

Sikora, L.J., and N.K. Enkiri. 2003. Availability of poultry litter compost P to fescue compared with triple super phosphate. *Soil Sci.* 168:192–199.

Sikora, L.J., and R.A.K. Szmidt. 2001. Nitrogen sources, mineralization rates, and nitrogen nutrition benefits to plants from composts, pp. 287–306. In P.J. Stoffella and B.A. Kahn (eds.), *Compost Utilization in Horticultural Cropping Systems*, CRC Press, Boca Raton, FL.

Sooby, J., J. Landeck, and M. Lipson. 2007. Soil: Microbial life, fertility management, and soil quality, pp. 20–33. In *National Organic Research Agenda*, Organic Farming Research Foundation, Santa Cruz, CA.

Spargo, J.T., G.K. Evanylo, and M.M. Alley. 2006. Repeated compost application effects on phosphorus runoff in the Virginia Piedmont. *J. Environ. Qual.* 35:2342–2351.

Sterrett, S.B., R.L. Chaney, C.W. Reynolds, F.D. Schales, and L.W. Douglas. 1982. Transplant quality and metal concentration in vegetable transplants grown in media containing sewage sludge compost. *HortScience* 17:920–922.

Stivers-Young, L. 1998. Growth, nitrogen accumulation, and weed suppression by fall cover crops following early harvest of vegetables. *HortScience* 33:60–63.

Sullivan, P. 2003. Overview of cover crops and green manures. Appropriate technology transfer for rural areas. Natl. Sustainable Agr. Information Ctr.

Treadwell, D., M.A. Alligood, C.A. Chase, and M. Bhan. 2008a. Soil nitrogen responses to increasing crop diversity and rotation in organic vegetable production systems. *HortScience* 43:1107 (abstr.).

Treadwell, D., W. Klassen, and M. Alligood. 2008b. Annual cover crops in Florida vegetable systems, Part 1. Objectives: why grow cover crops?. Univ. Florida, Inst. Food Agr. Sci. EDIS HS387.

Treadwell D., W. Klassen, and M. Alligood. 2008c. Annual cover crops in Florida vegetable systems, Part 2. Production. Univ. Florida, Inst. Food Agr. Sci. EDIS HS114.

Tyler, R. 2001. Compost filter berms and blankets take on the silt fence. *Biocycle* 42(1):41–46.

Tyson, S.C., and M.L. Cabrera. 1993. Nitrogen mineralization in soils amended with composted and uncomposted poultry litter. *J. Commun. Soil Sci. Plant Anal.* 24:2361–2374.

U. S. Environmental Protection Agency, 1994. A plain English guide to the EPA part 503 biosolids rule. EPA832-R-93-003. Sept. Washington, DC.

U. S. Environmental Protection Agency, 1995. A guide to the biosolids risk assessments for the EPA part 503 rule. EPA832-B-93-005. Sept. Washington, DC.

U. S. Environmental Protection Agency, 1999. Biosolids generation, use, and disposal in the United States. EPA503-R-99-009. Sept. Washington, DC.

Van Kessel, J.S., J.B., Reeves, and J.J. Meisings. 2000. Nitrogen and carbon mineralization of potential manure components. *J. Environ. Qual.* 29:1669–677.

VanTine, M. and S. Verlinden. 2003. Growing organic vegetable transplants. West Virginia Univ. Ext. Serv.

Wallace, P. 2006. Production of guidelines for using compost in crop production-A brief literature review, Project code ORG 0036. Waste Resources Action Programme, Banbury, UK. 24 April 2020. http://www.cre.ie/docs/Nitrogen%20Review.pdf

Yang, J., Z. He, Y. Yang, P.J. Stoffella, X.E. Yang, D.J. Banks, and S. Mishra. 2007. Use of amendments to reduce leaching of phosphate and other nutrients from a sandy soil in Florida. *Environ. Sci. Pollution Res.* 14:266–269.

Yuran, G.T., and H.C. Harrison. 1986. Effects of genotype and sludge on cadmium concentration in lettuce leaf tissue. *J. Amer. Soc. Hort. Sci.* 111:491–494.

Zhang, M., and Y.C. Li. 2003. Nutrient availability in a tomato production system amended with compost. *Acta Hort.* 614:787–797.

chapter 4

Organic compost in crop production

Monica Ozores-Hampton

Organic crop production has been one of the fast-growing segments of U.S. agriculture of the past decade (Organic Trade Association (OTA), 2019). The United States is the largest organic food market in the world, with a value of $47.9 billion in 2018. Organic food sales increased by 5.9% from last year, as compared to a 2.3% growth rate in the overall conventional food market. However, organic sales account only for over 4% of the total U.S. food sales. The organic food market is characterized by mergers, acquisitions and investments. Therefore, such activities have led to large operations at all levels of the chain-food supply. The public perception that improves human health and the environment has been the main driver. The consumer appeal is that the organic seal is guaranteed by U.S. Department of Agriculture (USDA) certification that monitors an official standard, which includes non-GMO, toxic pesticides or chemicals, dyes, or preservations. Also, a strong differential price between organic and conventional crop production that growers can be capitalize in the market.

The sales of organic fruits and vegetables amounted to $17.4 billion in 2018, accounting for 36.3% of U.S. fruit and vegetable sales (OTA, 2019). There is a diversity of crops that are being sold as organic in the conventional market, natural food market, specialty supermarket, farmers market, community certified agriculture, etc. Certified organic acreage and livestock have been expanding in the United States for many years, especially fruit and vegetable, dairy and poultry (USDA/Economic Research Service (ERS), 2020; USDA/National Agricultural Statistics Service (NASS), 2020).

In 2016, there were 186,178, 15,398 and 46,250 acres dedicated to certified organic vegetable, berries and fruit/nuts production, respectively (USDA/ERS, 2020; USDA/NASS, 2020).

California remained the leading state in certified organic fruit and vegetable production, with over 147,523 acres and 2,713 farms (Figure 4.1). The USDA/ERS has estimated a 30% differential in fruit and vegetable prices in favor of organic than conventional (USDA/ERS, 2020).

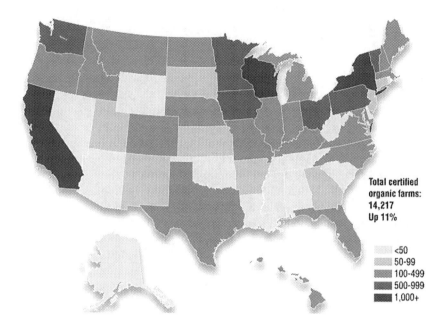

Figure 4.1 Number of certified organic farms by state, 2016. Source: Alexander, 2019; USDA/NASS, 2017.

Organic fruit and vegetable production in the United States must comply with the National Organic Program (NOP) standards (USDA/AMS, 2020). National Organic Program is a regulatory program in the U.S. Department of Agriculture (USDA) Agricultural Marketing Service and is responsible for developing the national standard in organic agricultural products. The USDA/NOP's responsibilities include accreditation of companies or organizations that certified organic farms and evaluation of the allowable agricultural inputs' materials. The definition of an organic production system under the NOP must be "managed under the act and regulated by the NOP to respond to site-specific conditions by integrating cultural, biological, and mechanical practices that foster cycling of resources, promote ecological balance, and conserve biodiversity" (USDA/AMS, 2020).

Compost products that meet the NOP standards are allowed to display the USDA organic seal for marketing purposes , except for operations with a gross income of $5,000 or less (Figure 4.2). The NOP defines compost as the product of a managed process through which microorganisms breakdown plant and animal materials into more available forms suitable for application to the soil (USDA/AMS, 2020). The NOP specified that compost must meet two criteria: (1) the initial carbon:nitrogen (C:N) ratio of the blended feedstocks should be between 25:1 and 40:1, and (2) the temperature must

Figure 4.2 U.S. Department of Agriculture Organic seal. Source: National Organic Program. USDA/AMS, 2020.

remain between 131° and 170° for 3 days in an in-vessel or static aerated pile or 15 days in windrows, which must be turned at least five times during this period (USDA/AMS, 2020). The NOP provides a National List (NL) of Allowed (natural) or Prohibited (synthetic) substances that can be used or not in organic production, processing or handling. Compost feedstocks such as YW, food waste, manure and other organic materials are allowable. Prohibited substances used as feedstocks during the composting process may include sewage sludge or biosolids, recycled wallboard, irradiated materials, GMOs (genetically modified organisms) or any synthetic material such as urea, except those listed in the NL of synthetic substances allowed in organic production (USDA/AMS, 2020). The application of sewage sludge or biosolids is prohibited, whether composted or uncomposted. Biosolids are solid, semisolid or liquid residue generated during the treatment of domestic sewage in treatment works. Sewage sludge includes, but is not limited to, domestic septage; scum or solids removed in primary, secondary or advanced wastewater treatment processes; and any material derived from sewage sludge.

The NOP has approved a variety of nonhazardous organic wastes that can be composted for land application to improve soil quality, which can be added in the NL (USDA/AMS, 2020). These include crop residues, animal manures, food waste, YW, lawn clippings and organic agricultural byproducts (Ozores-Hampton, 2012; Stoffella et al., 2014). The NOP regulations do not dictate zero tolerance for synthetic pesticide residues in compost feedstock such as yard and lawn clippings. Proper composting, as defined by the NOP, is expected to eliminate pathogens by heat and competition with other microorganisms. Therefore, no waiting period is required between compost application, in which the soil amendment does not come in contact with the edible product, and harvest time

(USDA/AMS, 2020). Commercial aerobic composting methods range from expensive operations, such as computerized in-vessel aerobic systems (turning, forced aeration and order control) or closed anaerobic systems, to inexpensive (manually or machine turned) windrow or static pile systems (Rynk and Richard, 2001).

Independent of the feedstock and composting methods, compost should be tested to help growers understand compost quality, nutrient content and potential effects in their production systems (Ozores-Hampton, 2017). The compost supplier should collect a composite sample of the compost and analyze it at intervals of every 20,000 tons of compost produced or every 3 months, whichever comes first. The recommended compost testing methodologies and sampling procedures should follow the "Test Methods for the Examination of Composting and Compost" that evaluate and verify that the compost meets physical, chemical and biological requirements (Ozores-Hampton, 2017; Ozores-Hampton et al., 2012). Also, before the compost is delivered to the grower, the supplier should provide the following documentation: (a) the compost meets federal and state health and safety regulations, (b) the composting process has met the time and temperature requirements of the NOP and (c) laboratory analysis that is less than 4 months old.

The organic composter can utilize the Organic Materials Review Institute (OMRI www.omri.org/) as a marketing tool for wholesale or retail compost for organic growers (OMRI, 2020). OMRI is an accredited 501(c)(3) nonprofit organization that operates independently of USDA/NOP and publishes the *OMRI Products List* (*OPL*). The *OPL* is comprised of brand-name products reviewed by OMRI and determined to be either allowed or allowed with restrictions in organic production, based on the NOP's U.S. organic standards. The *OPL* is updated regularly. Only products currently on the *OPL* are allowed to bear the OMRI Listed seal on labels and in advertising and promotions (Figure 4.3). In 2020,

Figure 4.3 Organic Materials Review Institute seal logo. Source: OMRI, 2020.

OMRI-approved products were 7,500, an increase of 13% from previous years (OMRI, 2020). Individual products can be OMRI Listed and not the compost facility, so if multiple compost types or a blend is produced by the facility, each product or blend must be OMRI approved. The cost of reviewing the products by OMRI will be equivalent to the gross sale and product type. Participation in the OMRI program is voluntary (OMRI, 2020). Thus, the *OPL* is not a comprehensive listing of products suitable for organic production, handling and processing. Also, a product absence from the *OPL* does not imply its failure to comply with the NOP program.

Certified compost as an organic product needs to meet the NOP standards which is a federal USDA organic program that has strict production and labeling requirements (USDA/AMS, 2020). The NOP rules require nothing artificial was incorporated into the production, processing, and handling of the compost. Once, the composter complies with the rules and regulations under the NOP, a USDA organic seal can be used in marketing the product. Additionally, the composter for a fee can list the organic compost in the OMRI/*OPL* and obtain an additional seal for labels, advertising and promotions.

References

Alexander, R. 2019. Is organic certification right for your compost? *BioCycle*, April 2019. https://www.biocycle.net/2019/03/11/organic-certification-right-compost/.

Organic Materials Review Institute. 2020. Welcome to Organic Materials Review Institute. 14 April 2020. https://www.omri.org/.

Organic Trade Association. 2019. U.S. organic industry survey. 15 July 2019. https://ota.com/resources/organic-industry-survey.

Ozores-Hampton, M. 2012. Developing a vegetable fertility program using organic amendments and inorganic fertilizers. *HortTechnology* 22:743--750.

Ozores-Hampton, M. 2017. Guidelines for assessing compost quality for safe and effective utilization in vegetable production. *HortTechnology* 27:150.

Ozores-Hampton, M., P. Roberts, and P.A. Stansly. 2012. Organic pepper production. In *Peppers: Botany, Production and Uses*, edited by V. Russo, 165–174. Cambridge, MA: CABI.

Rynk, R. and T.L. Richard. 2001. Commercial compost production systems. In *Compost Utilization in Horticultural Cropping Systems*, edited by P.J. Stoffella and B.A. Kahn, 51–93. Boca Raton, FL: Lewis Publishers.

Stoffella, P.J., Z.L. He, S.B. Wilson, M. Ozores-Hampton, and N.E. Roe. 2014. *Utilization of Composted Organic Wastes in Vegetable Production Systems*. Food & Fertilizer Technology Center, Technical Bulletins. Taipei, Taiwan. 12 April 2020. http://www.agnet.org/htmlarea_file/library/20110808105418/tb147.pdf.

U.S. Department of Agriculture/Agricultural Marketing Service. 2020. National Organic Program. 11 April 2020. https://www.ams.usda.gov/about-ams/programs-offices/national-organic-program.

U.S. Department of Agriculture/Economic Research Service. 2017. Organic market overview. 11 April 2020. https://www.ers.usda.gov/topics/natural-resources-environment/organic-agriculture/organic-market-overview.

U.S. Department of Agriculture/National Agriculture Service Statistics. 2017. Certified organic survey 2016 summary. 11 April 2017. https://downloads.usda.library.cornell.edu/usda-esmis/files/zg64tk92g/70795b52w/4m90dz33q/OrganicProduction-09-20-2017_correction.pdf.

chapter 5

Compost utilization in fruit production

Fernando Alferez

Contents

Fruit crop production is by nature a perennial monoculture. This fact implies using intensive production systems, lack of crop rotation and soil exploitation through the years. Together, these practices may affect soil composition and structure, resulting in less efficient root systems and diminishing productivity with time. In addition, some geographic areas in the tropics, subtropics and temperate climatic regions that are suitable for crop production can be low in soil organic matter. In both cases, soil amendments are an important component of a good management program. Composting the soil (organic amendments) is a way to increase soil fertility in the long term [1].

Agriculture must be more efficient and productive. For this, there is a need for better soil structure, increased soil fertility and decrease in soil erosion rate. These benefits can be achieved by adding composts to the soil [2]. Although compost alone is not sufficient to meet crop nutritional requirements [3], it may integrate with inorganic fertilizers, lowering costs and improving soil properties.

Also, there is an increasing trend among consumers in demanding organic fruit production with less input of inorganic fertilizers and pesticides, which are perceived by the public opinion as harmful or potentially hazardous [4]. In the past few years, an increasing body of work has been devoted to analyzing differences between conventional and organic fruit production, focusing on yield, quality and nutritional attributes, although the number of studies on compost influence on these parameters is still not abundant.

In this chapter, we will review the latest findings related to compost effects on tree crops, focusing on three different areas: agricultural sustainability, effects on soil structure and fertility and fruit quality as influenced by compost.

5.1 Compost and agriculture sustainability of tree crops

Sustainability is a concept that has gained momentum during the last couple of decades in agriculture. In the case of tree fruit, because of its perennial nature, sustainability involves obtaining a regular crop that produces revenue for the growers while maintaining a lower input (reduction in pesticide, water and fertilizer use) that warrants profitability. To achieve this, compost addition to the soil is a recommendable strategy. For a detailed review on this, see [5].

There are several areas of sustainability in agriculture in which compost may have a significant impact if adopted as part of management in tree crop orchards: compost may reduce soil erosion, improve soil structure and help to retain water. This results in a reduction of irrigation and fertilization needs. At the same time, compost may help to reduce greenhouse emissions.

Compost has the potential for offering long-term solutions for soil sustainability that conventional chemical fertilizers are not able to provide. This is especially true in the case of fruit crops due to their perennial nature. In addition, there is an increasing recognition of the necessity of carbon sequestration and immobilization. There is a general consensus in the scientific literature about the compost effects in increasing carbon sequestration, thus reducing the overall release of greenhouse gases into the atmosphere [6]. Efforts in carbon sequestration are increasing since agricultural soils are viewed as sites to immobilize organic carbon, thus counteracting the increase in emission of greenhouse gasses. In this regard, in a long-term study spanning 8 years of data collection, Regni et al. [1] showed that long-term treatments of an olive (*Olea europaea* L.) orchard, with compost derived from olive mill wastes and tree pruning materials, positively affected tree growth and olive fruit yield and represented an efficient strategy to increase soil fertility and carbon sequestration efficiency. At the same time, sustainability implies to achieve good cost-benefit balance. This is of paramount importance in organic agriculture, which presents specific challenges such as nitrogen (N) supply because organic fertilizers are expensive and release of N can be slow and unpredictable. In these situations, the addition of compost in combination with organic mulches has been proven to be more effective in increasing soil (N) availability in organic apple (*Malus domestica* Borkh.) tree orchards [7].

Compost is very effective in water retention of tree orchards. Compost mulches allow rainfall to penetrate in the soil and reach the root system. At the same time, they reduce water loss from evaporation. All in all, stress is reduced, and growth is improved, resulting in better fruit size and more yield [8].

Reduction of pesticide use is an increasing demand in modern society. To achieve this, integrated pest and disease management employing composts to improve soil health has been proven effective [5]. There is some evidence that compost and compost derivatives, such as compost teas, can be useful in reducing disease in fruit trees [9], and there is no consensus in studies. In general, compost rates of at least 20% (v/v) are required for consistent disease suppression [10]. It seems clear that the effectiveness of the compost treatment depends on the cultivar and pathogen. For instance, compost tea treatment has not been useful in controlling *Ventura inequalis* (the causal agent of apple scab) in apple, but it controls *Monilinia laxa* growth in cherry (*Prunus avium.* L.) blossom, thus reducing brown rot [11]. A comprehensive list of treatments, cultivars and disease is reported by Scheurell and Mahaffee [9]. In any case, compost disease suppression is a transient effect that depends on different groups of organisms [12], and compost effects are better if combined with conventional fumigation, especially in cases where new plantings have to coexist with older trees in a soil compromised by root pathogens already existing. In this sense, Braun et al. [13] showed that Honeycrisp apple trees treated with a combination of fumigation and compost in a multiyear study performed better than non-compost treated.

Interestingly, compost added to the soil may have a beneficial effect as it promotes biodegradation of pesticides. Pesticide persistence is a serious concern because recalcitrant pesticides may harm the environment through leaching into aquifers and altering the microbial community and biological processes in the soil [14]. As the microbial community changes with the addition of compost, degradation of pesticides may occur faster. In this sense, newly added, young compost can absorb and degrade pesticides, due to greater carbon sources for microorganisms to metabolize and remove pesticides [15].

5.2   Soil structure and fertility

Addition of compost to the soil increases organic carbon [2,16] and enhances soil structure and fertility while decreasing erosion [2]. This is especially important in semi-arid regions with high potential for tree crop production, such as the Mediterranean climatic areas [17]. In a long-term study, Regni et al. [1] found that besides increasing organic carbon in the soil, compost improved vegetative growth and fruit of olive trees. These researchers attributed this result to a direct effect of the treatment on the amount of total N and availability of phosphorus and potassium in the soils. However, as they noted in their study, other factors such as water retention and microbial biomass activity are also improved by compost addition, and in turn, may contribute to the improvement of tree health and performance.

For adequate tree growth and yield, N must be supplied at the right times. In apples, in-season N uptake and reserve N stored in perennial tissues from previous growth cycles are vital for the tree nutrition and growth. Remobilization of these reserves is the main driver of early-season fruit and leaf growth, but partitioning of in-season N uptake is affected by the fertilizer timing [7]. These researchers found that vigorous tree growth after compost treatment allowed young apple trees to establish quickly and yield well in the first harvest year [7].

Compost addition has been proven to be effective also for amending specific nutrient deficiencies in poor soils. For instance, in alkaline calcareous soils in which iron (Fe) may not be available to the trees, compost may be effective in reducing this deficiency effects. In pear (*Pyrus communis* L.) tree orchards in the Mediterranean, it has been shown that compost addition improved Fe nutrition and prevented Fe chlorosis while increasing fruit quality in terms of size and soluble sugar concentration, and in addition, increased soil carbon biomass [18].

In citrus, there is interest in the use of compost to improve soil quality, especially in areas with poor soils and where the disease Huanglongbing (HLB, or citrus greening) is present. This is the case in Florida. In Florida, virtually all citrus acreage is affected by HLB, which causes a premature tree to decline and decrease fruit yield and quality. Currently, HLB is the most devastating citrus disease worldwide. One of the main effects of the disease is the reduction in root density [19]. Florida soils are generally sandy and with low content of organic matter. Compost has been shown to amend soil quality in sandy areas with an increase in available minerals [20], and it has been proposed as a complement to foliar nutritional programs by enhancing the root environment and by providing a slow-release carbon source for active soil microorganisms. In this case, the best effects are observed when compost is applied at a shallow depth to avoid root damage in already compromised root systems due to disease [21]. In this sense, compost derivatives, such as humic and fulvic acid, were able to increase root health and fibrous root density when applied to HLB-infected trees, in which as much as 80% of the root system may be lost [22].

5.3   Compost and tree fruit quality

It has been proposed that changes caused by compost in soil chemical and physical characteristics may result in the increase of beneficial microorganisms and nutrient availability and uptake, thus favoring plant and fruit growth as well as yield and nutritional attributes of these fruit [23] (Wang, 2007). However, this is still under controversy, since the effect of compost seems to vary greatly depending on the compounds studied, the initial soil fertility and the fruit developmental or ripening stage. In addition, fruit variety seems to respond differently. In summary, there exist enormous variations in compost effects reported on fruit quality. Unfortunately, deep studies on the molecular, biochemical and physiological mechanisms involving the increase of quality by compost addition are not abundant. Here, we review some studies published in the past years.

For example, in peaches (*Prunus persica* (L.) Batsch.) and pears culti-vated with compost, polyphenol content and polyphenol oxidase activity increased in parallel when compared to conventional cultivated trees, and other compounds such as ascorbic acid and α-tocopherol also increased [24]. In cherries, a multiyear study has shown some improvement in fruit quality as related to total soluble solids [25]. In guava (*Psidium guajava* L.), it has been shown that composting may induce more vegetative growth and fruit yield, whereas fruit quality is comparable to conventional farm-ing [1]. These results were similar to those observed by Bastos de Melo et al. [26] in other study conducted with peaches, in which quality attri-butes (soluble solids, acidity and fruit color) were not affected but the yield increased. Other studies have shown that compost addition may have an effect, promoting fruit development and maturation. This may allow growers to advance harvest for targeting market opportunities and increasing profit. For example, compost application to a conventionally managed pear orchard resulted in more attractive skin color, sweetness and reduced stone volume [27]. These authors suggest that the increased leaf nutrients after composting could change the postharvest behavior of fruit and accelerate fruit maturation; furthermore, since organic fertiliza-tion provides less N due to the slow mineralization rate, this may, in con-sequence, improve fruit quality because smaller fruit would have more concentration in calcium. In other cases, adding compost in combination with biochars did not result in benefit for the grove or in an increase of fruit quality. This is especially true in the case of fertile soils, as shown by von Glisczynski et al. [28] in German apple orchards. In general, posi-tive effects are found in short-term studies, whereas these effects are less clear in long-term studies. In this sense, Toselli et al. [29] did not find an increase in fruit quality in an 8-year study conducted in Italy with peaches. Similarly, Khorram et al. [30] found that application of biochar and compost in an apple orchard, separately or in combination, improved some soil properties during the first 2 years of application and that vegeta-tive growth increased but not fruit yield, weight or starch pattern index as productivity indices. Other types of compost, such as compost tea, can be applied foliar and may increase yield and quality in pears, especially if combined with antioxidants such as ascorbic acid [31].

References

1. Regni, L., Nasini, L., Ilarioni, L., Brunori, A., Massaccesi, L., Agnelli, A., and Proietti, P. 2017. Long term amendment with fresh and composted solid olive mill waste on olive grove affects carbon sequestration by prunings, fruits, and soil. *Frontiers in Plant Science* 7:2042.
2. Sánchez-Monedero, M.A., Cayuela, M.L., Mondini, C., Serramiá, N., and Roig, A. 2008. Potential of olive mill wastes for soil C sequestration. *Waste Management* 28:767–773.
3. Pinamonti, F., and Sicher, L. 2001. Compost utilization in fruit production systems. In *Compost Utilization in Horticultural Cropping Systems*. Lewis Publishers: New York, NY, USA, 2001; pp. 177–200.
4. Athani, S.I., Ustad, A.I., Prabhuraj, H.S., Swamy, G.S.K., Patil, P.B., and Kotikal, Y.K. 2007. Influence of vermi-compost on growth, fruit yield and quality of guava cv. sardar. *Proceedings of Ist IS on Guava*. Edited by G. Singh et al. *Acta Horticulturae* 735, ISHS 2007.
5. Litterick, A.M., Harrier, L., Wallace, P., Watson, C.A., and Wood, M. 2004. The role of uncomposted materials, composts, manures, and compost extracts in reducing pest and disease incidence and severity in sustainable temperate agricultural and horticultural crop production- A review. *Critical Reviews in Plant Sciences* 23:453–479.
6. Walker, P., Williams, D., and Waliczek, T. 2006. An analysis of the horticulture industry as a potential value-added market for compost. *Compost Science & Utilization* 14:23–31.
7. TerAvest, D., Smith, J.L., Carpenter-Boggs, L., Hoagland, L., Granatstein, D., and Reganold, J.P. 2010. Influence of orchard floor management and compost application timing on nitrogen partitioning in apple trees. *Horticultural science* 45:637–642.
8. Buckerfield, J.C., and Webster, K.A. 2000. *Composted "green organics" for water conservation and weed control*. Urrbrae, South Australia: CSIRO Land & Water.
9. Scheurell, S., and Mahaffee, W. 2002. Compost tea: Principles and prospects for plant disease control. *Compost Science and Utilization* 10:313–338.
10. Noble, R., and Coventry, E. 2005. Suppression of soil-borne plant diseases with composts: A review. *Biocontrol Science and Technology* 15:3–20.
11. Pscheidt, J., and Witting, H. 1996. *Fruit and Ornamental Disease Management Testing Program*. Corvallis, OR: Extension Plant Pathology, O.S.U.
12. Boulter-Bitzer, J.L., Trevors, J.T., and Boland, G.J. 2006. A polyphasic approach for assessing maturity and stability in compost intended for suppression of plant pathogens. *Applied Soil Ecology* 34:65–81.
13. Braun, P.G., Fuller, K.D., McRae, K., and Fillmore, S.A.E. 2010. Response of "Honeycrisp" apple trees to combinations of pre-plant fumigation, deep ripping, and hog manure compost incorporation in a soil with replant disease. *Horticultural science* 45:1702–1707.
14. Cardinali, A., Otto, S., Vischetti, C., Brown, C., and Zanin, G. 2010. Effect of pesticide inoculation, duration of composting, and degradation time on the content of compost fatty acids, quantified using two methods. *Applied and Environmental Microbiology* 76:6600–6606.

15. Wu, L., and Ma, L.W. 2000. Relationship between compost stability and extractable organic carbon. *Journal of Environmental Quality* 31:1323–1328.
16. Russo, G., Vivaldi, G.A., De Gennaro, B., and Camposeo, S. 2015. Environmental sustainability of different soil management techniques in a high-density olive orchard. *Journal of Cleaner Production* 107:498–508.
17. Martínez-Mena, M., Alvarez-Rogel, J., Castillo,V., and Albaladejo, J. 2002. Organic carbon and nitrogen losses influenced by vegetation removal in a semiarid Mediterranean soil. *Biogeochemistry* 6:309–321.
18. Sorrenti, G., Toselli, M., and Marangoni, B. 2012. Use of compost to manage Fe nutrition of pear trees grown in calcareous soil. *Scientia Horticulturae* 136:87–94.
19. Johnson, E.G., Wu, J., Bright, D.B., Graham, J.H. 2014. Association of "Candidatus Liberibacter asiaticus" root infection, but not phloem plugging with root loss on huanglongbing-affected trees prior to appearance of foliar symptoms. *Plant Pathology* 63:290–298.
20. Hoffman, J. 2017. Soil improvement with organic mulch. *Citrus Industry* 98: 26–28.
21. Ozores-Hampton, M., Adair, R., and Stansly, P. 2015. Using compost in citrus. *Citrus Industry* 96:8–11.
22. Brown, J., and Kamberg, C. 2016. Rebuilding the foundation. *Citrus Industry* 97:26–30.
23. Wang, S.Y. 2007. Effect of pre-harvest conditions on antioxidant capacity in fruits. *Proceedings of the IVth IC on MQUIC.* Edited by A.C. Purvis et al. *Acta Horticulturae* 712, ISHS.
24. Carbonaro, M., Mattera, M., Nicoli, S., Bergamo, P., and Cappelloni, M. 2002. Modulation of antioxidant compounds in organic vs conventional fruit (peach, Prunus persica L., and pear, persica L., and pear, Pyrus communis L.). *Journal of Agricultural and Food Chemistry* 50:5458–5462.
25. McPhee, W.J. 2012. *Evaluation of Compost on Crop Production, Tree Growth, Fruit Quality and Soil Fertility.* RDOS Compost Evaluation Project Sponsored by: Southern Interior Development Initiative Trust, Regional District of Okanagan-Similkameen, Okanagan Kootenay Cherry Growers Association.
26. Bastos de Melo, G.W., Sete, P.B., Ambrosini, V.G., Freitas, R.F., Basso, A., and Brunetto, G. 2016. Nutritional status, yield and composition of peach fruit subjected to the application of organic compost. *Acta Scientarum, Agronomy* 38:103–109.
27. Lee, J.-A., Kim, W.-S., and Choi, H.-S. 2009. Effect on fruit quality of 2-year compost application in a conventionally managed pear orchard. *Korean Journal of Food Preservation* 16(3):317–320.
28. Von Glisczynski, F., Sandhage-Hofmann, A.W., and Pude, R. 2016. Biochar-compost substrates do not promote growth and fruit quality of a replanted German apple orchard with fertile Haplic Luvisol soils. *Scientia Horticulturae* 213:110–114.
29. Toselli, M., Baldi, E., Marcolini, G., Quarteri, M., Sorrenti, G., Marangoni, B., and Innocenti, A. 2012. Long term (8 years) effect of mineral and organic fertilizations on peach yield and nutritional status. *Proceedings of 7th International Peach Symposium*, Edited by J. Girona and J. Marsal. *Acta Horticulturae* 962:371–378. ISHS.

30. Khorram, M.S., Zhang, G., Fatemi, A., Kiefer, R., Maddah, K., Baqar, M., Zakaria, M.P., and Li, G. 2019. Impact of biochar and compost amendment on soil quality, growth and yield of a replanted apple orchard in a 4-year field study. *Journal of the Science of Food and Agriculture* 99:1862–1869.
31. Fayek, M.A., Fayed, T.A., El-Fakhrani, E.M., and Sayed, S.N. 2014. Yield and fruit quality of "Le-conte" pear trees as affected by compost tea and some antioxidants applications. *Journal of Horticultural Science & Ornamental Plants* 6:01–08.

chapter 6

Compost utilization in vegetable production

Monica Ozores-Hampton

Contents

The United States ranks third worldwide in vegetable production with 2.5 million acres harvest (753 million cwt), representing a farm value of $12.9 billion in 2018. More than 50 vegetable commodities are grown commercially in the United States (U.S. Department of Agriculture (USDA), 2019). The top three vegetables in acres harvest were sweet corn (*Zea mays* L.), tomatoes (*Solanum lycopersicum* L.) and snap beans (*Phaseolus vulgaris*), and in production are tomatoes, sweet corn and onions (*Allium cepa* L.), which together accounted for 56% of all the vegetables. Tomatoes and romaine lettuce (*Lactuca sativa* L.) had the highest value, accounting for 31% of the total value. California continues to be the leading state in area harvested, production and value in the United States (U.S. Department of Agriculture (USDA), 2019).

In vegetable, large-seeded crops such as sweet corn, snap bean, English pea (*Pisum sativum*), southern pea (*Vigna unguiculata*) or okra (*Abelmoschus esculentus*) are typically directly seeded in open beds. Whereas transplants (grown in a greenhouse in Styrofoam trays) are usually used for the establishment of small-seeded crops such as tomato, bell pepper (*Capsicum annuum* L.), eggplant (*Solanum melongena.* L.), watermelon (*Citrullus lanatus* (Thunb.) Matsum. & Nakai) triploid (seedless), strawberry (*Fragaria vesca* ssp. *americana* (Porter) Staudt), melons, cucumber (*Cucumis sativus* L.) or summer or zucchini squash (*Cucurbita pepo*) may be directly seeded on raised beds. Raised beds may be covered with a polyethylene mulch or remain uncovered. Polyethylene-mulched beds may be used to grow one, two (double cropping) or three (triple cropping) crops. Plasticulture production systems beside polyethylene include mulch raised beds, fumigation, irrigation and soluble fertilizer application, and open bed production includes herbicides, irrigation and soluble fertilizer application (Ozores-Hampton et al., 2015).

6.1   U.S. government, state and county/city compost regulation in vegetable production

There are no U.S. government restrictions on how and when compost can be used in vegetable production, except for compost derived from sewage sludge or biosolids (BS) (Ozores-Hampton and Peach, 2002; U.S. Environmental Protection Agency (USEPA), 1994; 1995, 1999). To eliminate or reduce human and plant pathogens, nematodes and weeds, the temperature during the composting process (thermophilic stage) must have an average of 131°F for 3 days in an in-vessel or static aerated pile or first 15 days in windrow composting methods and be turned at least five times during this period, according to the Clean Water Act 40 CFR Part 503 (USEPA, 1994, 1995, 1999). Also, the Clean Water Act 40 CFR Part 503 was incorporated in the National Organic Program (NOP) standards (USDA/AMS, 2020). Therefore, organic producers of vegetable crops will not require a waiting period between compost application and planting or seeding as long as the compost does not come into contact with the edible product and harvest time in contrast to raw manure application (USDA/AMS, 2020). U.S states, counties and cities may impose more strict regulation than the U.S. government with the purpose to address local, unique environmental conditions. All compost that contains regulated feedstocks must meet national, state and/or local safety standards to be marketable.

6.2   Sources of compost in vegetable production

Compost to be used in vegetable production can be produced from a variety of feedstocks produced by urban populations such as municipal solid waste (MSW); yard waste (YW); food wastes from restaurants, grocery stores and institutions; wood wastes from construction and/or demolition; wastewater (from water treatment plants); and BS (Ozores-Hampton, 2017, 2012). Agriculture produces other organic wastes that can be composted: poultry, dairy, horse, feedlot and swine manures;

wastes from food processing plants; and spoiled feeds, harvest wastes and mushroom media.

6.3   Compost quality guidelines in vegetable production

There are no uniform standards on compost use in vegetable production (Ozores-Hampton, 2017). The lack of a uniform standard that limits the development of compost use guidelines is due to the diversity and variability of a number of factors such as feedstocks (animal manures, food waste, yard trimming waste, agricultural by-products, etc.), composting methods (windrow, static passive or aerated piles, in vessels, etc.), application rates, application time (before or at planting), application methods (broadcast or band), crop type (fruit, leafy, stems, tubers or roots), time of the year (fall, winter or spring) and application alone or combined with other organic or inorganic nutrient sources (Ozores-Hampton, 2017; Stoffella et al., 2014).

Compost quality guidelines are still limited and non-comprehensive in addressing the compost potential once is incorporated into the soil. These physical, chemical and biological compost quality guidelines can promote the positive effects of compost on soil/crops and minimize the negative impacts (environmental, crop production and growth and others) on organic or conventional vegetable production. Parameters relevant to consider when using compost in vegetable production are (Table 6.1): bulk density, moisture, organic matter, particle size, physical contaminants (inert materials), pH, electrical conductivity (EC; soluble salts), stability, C:N ratio, N–P–K content, heavy metals, maturity (growth screening), weeds free and human pathogen (Ozores-Hampton, 2017). These physical, chemical and biological compost quality guidelines will promote the positive effects of compost on soil/crops and minimize the negative impacts (environmental, crop production and growth and others) on vegetable production.

6.4   How and when to incorporate compost in vegetable production

Compost may be applied using a traditional manure spreader (flail/rear discharge or side discharge) or other specialized equipment. Compost can be applied throughout an entire field (broadcast) but higher performance is associated with band application only in the rows or bed (Ozores-Hampton et al., 1998a; Ozores-Hampton, 2012). The product should be uniformly applied to the surface and then incorporated to an approximate

Table 6.1 Optimal compost physical, chemical and biological properties range for use in vegetable production and other production systems (Ozores-Hampton, 2017)

Parameter (unit)	Optimal range	TMECC methods no.
Physical		
Bulk density (lb/yard³ wet basis)	740–980	3.03
Moisture (%)	30 (dry)–60 (wet)	–
Organic matter (%)	40–60	5.07-A
Particle size	98% pass through 3/4-inch screen or smaller than 1 inch	2.02-B
Physical contaminants (%)	<2%	3.08-A
Chemical		
pH	5.0–8.0	4.11-A
Electrical conductivity (mmhoᵘ/cm)	<6	4.10-A
Stability (Carbon dioxide (CO_2) evolution rate or oxygen consumption)	CO_2-C /unit volatile solid (VS)/day as <2 = very stable; 2–8 = stable and >8 = unstable. Oxygen (O_2) uptake O_2/ VS/h as <0.5 = very stable; 0.5–1.5 = stable and >1.5 = not stable	5.08-B
Solvita maturity test (Woods End Research Laboratory, Mt Vernon, ME)	≥6	–
C/N (carbon:nitrogen)	10–25	4.01 and 4.02
Nitrogen (%)	0.5–6.0	4.02
Phosphorous (%)	0.2–3.0	4.03
Potassium (%)	0.10–3.5	4.04
Heavy metals	Meet or exceed USEPA Class A standard, 40 CFR § 503.13 or DEP 62-709	

(Continued)

Table 6.1 (Continued) Optimal compost physical, chemical and biological properties range for use in vegetable production and other production systems (Ozores-Hampton, 2017)

Parameter (unit)	Optimal range	TMECC methods no.
Arsenic (As) (ppm[u])	<41	4.06-As
Cadmium (Cd) (ppm)	<15 (DEP)	4.06-Cd
Copper (Cu) (ppm)	<450 (DEP)	4.06-Cu
Lead (Pb) (ppm)	<300	4.06-Pb
Mercury (Hg) (ppm)	<17	4.06-Hg
Molybdenum (Mo) (ppm)	<75	4.06-Mo
Nickel (Ni) (ppm)	<50 (DEP)	4.06-Ni
Selenium (Se) (ppm)	<100	4.06-Se
Zinc (Zn) (ppm)	<900 (DEP)	4.06-Zn
Biological		
Maturity (Seed emergence and seedling vigor)	>80% relative to positive control	5.05-A
Weed-free	No or very low weed seeds	–
Pathogen	Meet or exceed USEPA Class A standard, 40 CFR § 503.32(a)	
Fecal coliform (MPN/g TS)[u]	<1,000	7.01
Salmonella (MPN/4 g)	<3	7.02

TMECC = Test Methods for the Examination of Composting and Compost;

MPN = most probable number;

TS = total solids.

depth of 5–6 inches using a rototiller, disk or mole board plow or other tillage equipment.

6.5   Compost effects in vegetable production

The horticultural industry is the primary consumer of compost in the world. Therefore, the composting process represents the most widely used recycling organic waste technology (Stentiford and Sanchez-Monedero, 2016). Growers can use compost as a soil conditioner or as a nutrient source to supplement the fertility program in vegetable production. Mature and stable compost utilization in vegetable production as a soil amendment can improve plant growth/development and yields/quality by improving the soil quality with the addition of organic matter (Aggelides and Londra, 2000; Bulluck et al., 2002; Ghaly and Alkoaik, 2010; Montemurro et al., 2005; Ozores et al., 2011, 2012; Pane et al., 2015; Setyowati et al., 2014; Stoffella et al., 2014). A soil amendment compost can be applied to improve soil quality such as soil physical properties (water-holding capacity, soil structure and bulk density), soil chemical properties (cation exchange capacity and plant nutrient availability) and soil biological properties (microbial activity) (Bulluck et al., 2002; Bhattacharyya et al., 2003; Guerrero et al., 2000; Li et al., 2010; Ozores-Hampton, 2012; Ozores et al., 2011, 2012; Salazar et al., 2012; Stoffella et al., 2014). Similarly, compost application can reduce nutrient leaching and the potential impact of the risk of nutrient in the environment, especially in water bodies (Li et al., 2010; Massri and Labban, 2014; Stoffella and Graetz, 2000; Ozores-Hampton et al., 2011).

Most often compost is recommended as a soil conditioner, depending on the feedstocks used to make the compost and the quality of the final product, which may contain significant nutrient amounts. Compost as a nutrient source offers a slow supply of macro and micronutrients essential to plant growth and development, especially after consecutive application. Hence, testing compost quality and nutrient composition may be relevant, especially when compost is used to supplement the fertility program (Ozores-Hampton, 2012, 2017). However, limited studies utilized compost as partial substitute inorganic synthetic fertilizer in vegetable production (Rosen and Bierman, 2016; Stoffella et al., 2014). Depending on the application rates, the amount of phosphorus (P), zinc (Zn), manganese (Mn), boron (B), iron (Fe) and molybdenum (Mo) can be significant and need to be quantified in the fertility program (Ozores-Hampton, 2012). Tomatoes grown with sugarcane compost at the rate of 98 ton/acre, with a 50% nitrogen (N) synthetic fertilizer standard rate of 300 lb/acre, improved fruit yields and fruit size when compared to the 100% standard growers' N fertilizer program (Stoffella and Greatz, 2000). Similarly, MSW compost at the rate of 34 ton/acre (dry weight basis) combined with inorganic synthetic fertilizer increased tomato marketable yields, probably due to higher extractable soil P, potassium (K), Zn, Fe, B and Mo at the root zone than inorganic synthetic fertilizer application (Zhang et al., 2003). Therefore, blending compost and inorganic synthetic fertilizer have been more effective in producing positive crop responses than separate application of either material alone (Hernández et al., 2014).

Amending soils with mature and stable composted materials has been reported to increase vegetable vigor, yields and quality of cucumber (*Cucumis sativus* L.), bell peppers (*Capsicum annuum* L.), eggplant (*Solanum melongena* L), lettuce, muskmelon (*Cucumis melo* L.), potatoes (*Solanum tuberosum* L.), tomatoes, squash (*Cucurbita pepo* L.) and watermelons (*Citrullus lanatus* (Thunb.) Matsum. & Nakai) (Tables 6.2, 6.3 and 6.4) (Mehdizadeh et al., 2013; Ozores-Hampton, 2012; Ozores et al., 2011, 2012; Stoffella et al., 2014; Stoffella and Graetz, 2000; Taiwo et al., 2007; Zhang et al., 2003). Vegetables have been produced using a wide range of compost application rates of 5–70 tons/acre. Lower rates of compost are typically being used as "maintenance" applications. Appropriate compost application rates will be influenced by existing soil conditions, compost characteristics and the crop nutrient requirements. Yard waste compost application at the rate of 33 ton/acre enhanced soil organic matter, and 100 ton/hectare increased the soil moisture content in tomato production (Loper et al., 2010; Shober and Vallad, 2012). Application rates of 5.0 ton/acre MSW compost produced 58% higher tomato yields than non-amended soils (Maynard, 1995). Similarly, composted DM, MSW and paper mill BS increased yield, tuber dry matter percent and tuber nitrate content potato (Barmaki et al., 2008; Ghaly and Alkoaik, 2010; Lalande et al., 2003). Composted DM applications enhanced crop growth and marketable yields of cucumber and heads per plant, head size and marketable yield of broccoli (*Brassica oleracea* L.) (Nair and Ngouajio, 2010; Roe and Gerald, 2010). In sandy and calcareous soil, MSW compost application rates of 40 tons/acre resulted in crop yield increases for bean and watermelon (Ozores-Hampton and Bryan, 1993b; Obreza and Reeder, 1994). However, no yield increases were observed on bell pepper marketable yield between BS compost and non-amended soils (Ozores-Hampton and Stansly, 2005). Similarly, no yield effect was reported on soils treated with DM/food waste, brewery waste or MSW compost bell peppers (Clark et al., 2000; Aram and Rangarajam, 2005).

Unstable compost consistently caused "N-immobilization" or in which available forms of inorganic N are converted to unavailable organic N and inhibited growth of vegetable crops such as beans (Kostewicz, 1993), corn (Gallaher and McSorley, 1994a,b), peppers, tomatoes and squash (Bryan et al., 1995). When immature and unstable compost is applied and a crop is planted immediately, growth inhibition and stunting may be visible for 40–60 days. When using compost with C:N ratios higher than 25 or 30,

Table 6.2 The effects of yard trimming (YW) compost, rate and soil type on vegetable crop yield responses

Crop type	Compost type	Compost rate (ton/acre)		Crop response	Reference
Corn	YW	0, 60 and 120 (1 year)	Sandy	No response N-immobilization	Gallaher and McSorley, 1994a
Corn	YW	240, 300 and 360 (3 year)	Sandy	Increased yields	Gallaher and McSorley, 1994a
Pepper	YW/BS	0 and 60	Sandy	Increased yields	Stoffella, 1995
Cucumber	YW/BS	0 and 60	Sandy	Increased yields in residual compost	Stoffella, 1995
Pole beans	YW	0, 25, 50 and 100 (1 year)	Sandy	100 ton/acre appears to be optimum rate with adequate fertilizer	Kostewicz, 1993
Pole beans	YW	0, 25, 50 and 100 (2 year)	Sandy	Immature compost caused "N-immobilization" and reduced yield. Strong residual effects.	Kostewicz, 1993
Okra	YW	0, 10 and 20	Sandy	No response	Kostewicz and Roe, 1991
Peas	YW	0, 10 and 20	Sandy	No response	Kostewics and Roe, 1991
Sweet corn	YW/BS	0, 10, 20 and 40	-	Increased yield	Jackson et al., 2013
Tomatoes/Peppers/Cucumbers	YW/BS	0 and 60	Sandy	Increased yield	Stoffella, 1995

BS = biosolids.

Table 6.3 The effects of municipal solid waste (MSW) compost, rate and soil type on vegetable crop yield responses

Crop type	Compost type	Compost rate (ton/acre)		Crop response	Reference
Bush beans	MSW	0, 40 and 60	Calcareous	No yield response with fertilizer; however, yield was increased with lower fertilizer rate	Ozores-Hampton and Bryan, 1993b
Bush beans	MSW	0, 36 and 72	Calcareous	Increased marketable yield by 25%	Ozores-Hampton and Bryan, 1994a
Black-eyed peas	MSW	0, 3 and 9	Sandy	At low and high N rate, compost increased yield	Bryan and Lance, 1991
Broccoli	MSW	0, 3, 6 and 12	Sandy	No response to mature compost	Roe et al., 1990
Corn	MSW	10, 20 and 30	-	Increased yield	Mkhabela and Warman, 2005
Eggplant	MSW	0, 40 and 60	Sandy	Increased yield with mature compost	Ozores-Hampton and Bryan, 1993a
Peppers	MSW/DM	0, 50 and 100	Sandy	Unmulched plot had the lowest yield, composts were next, and plastic mulch was the highest. There were N differences between compost mulch rates	Roe et al., 1992
Peppers	MSW	0, 40 and 60	Sandy	Increased yield with mature compost	Ozores-Hampton and Bryan, 1993a
Peppers	MSW/DM	6, 18 and 54	Sandy	Decreased yield with compost mulch compared to plastic mulch	Roe et al., 1993a
Peppers	MSW	0, 67 and 134	Calcareous	No effects	Clark et al., 2000
Potato	MSW	10, 19 and 29	-	Increased yield	Ghaly and Alkoaik, 2010

(Continued)

Table 6.3 (Continued) The effects of municipal solid waste (MSW) compost, rate and soil type on vegetable crop yield responses

Crop type	Compost type	Compost rate (ton/acre)		Crop response	Reference
Southern peas	MSW	0, 36 and 72	Calcareous	Increased marketable yield by 100%	Ozores-Hampton and Bryan, 1994a
Squash	MSW/BS	0, 15 and 30 (year 1)	Calcareous	Immature compost from previous season did not reduce yield	Bryan et al., 1995
Squash	MSW	0, 30 and 60 (year 1)	Calcareous	Increased yield with mature compost	Bryan et al., 1995
Squash	MSW/DM	0, 100 and 150	Sandy	Yield was higher in compost mulch than in plastic, due to disease problems in plastic	Roe et al., 1993a
Squash	MSW	10, 21.5 and 32.1	–	Increased yield	Ghaly and Alkoaik, 2010
Squash	MSW/BS	0 and 11	Calcareous	No response to mature compost	Ozores-Hampton et al., 1994b
Squash	MSW	0 and 21	Calcareous	No response to mature compost	Ozores-Hampton et al., 1994b
Tomato	MSW	0 and 33.4	–	Increased yield	Zhang et al., 2003
Tomato	MSW/BS	0, 6 and 12 (year 1)	Sandy	Increased yield with mature compost	Obreza and Reeder, 1994
Tomato	MSW/BS	0, 12 and 24 (year 2)	Sandy	Decreased yield with immature compost	Obreza, 1995
Tomato	MSW	0, 33 and 50 (year 1)	Sandy	Decreased yield with immature compost	Obreza and Reeder, 1994
Tomato	MSW	0, 66 and 100 (year 2)	Sandy	No response to mature compost	Obreza, 1995

(Continued)

Table 6.3 (Continued) The effects of municipal solid waste (MSW) compost, rate and soil type on vegetable crop yield responses

Crop type	Compost type	Compost rate (ton/acre)		Crop response	Reference
Tomato	MSW	0, 30 and 60	-	Increased yield	Clark et al., 2000
Tomato	MSW/BS	0, 15 and 30 (year 1)	Calcareous	Decreased yield with "immature compost"	Bryan et al., 1995
Tomato	MSW/BS	0, 30 and 60 (year 2)	Calcareous	Decreased yield with immature compost from previous year, but yield increased with mature compost	Bryan et al., 1995
Tomato	MSW	0, 60 and 120 (year 1)	Calcareous	Increased yield with mature compost	Bryan et al., 1995
Tomato	MSW/BS	0 and 11	Calcareous	No response to mature compost	Ozores-Hampton et al., 1994b
Tomato	MSW	0 and 21	Calcareous	No response to mature compost	Ozores-Hampton et al., 1994b
Watermelon	MSW/BS	0, 6 and 12 (year 1)	Sandy	No response to mature compost	Obreza and Reeder, 1994
Watermelon	MSW/BS	0, 12 and 24 (year 2)	Sandy	No response to mature compost	Obreza, 1995
Watermelon	MSW	0, 33 and 50 (year 1)	Sandy	Increased yield of 59%	Obreza and Reeder, 1994
Watermelon	MSW	0, 33 and 50 (year 2)	Sandy	No response to mature compost	Obreza, 1995

BS = biosolids; DM = dairy manure.

Table 6.4 The effect of compost feedstock source type, rate and soil types on vegetable crop yield responses

Crop type	Compost type	Compost rate (ton/acre)	Crop response	Reference
Bell pepper	DM/FW, brewery waste solids	0, 17.8 and 35.7	No effect	Aram and Rangarajam, 2005
Bell pepper	BS/DM/YW/MSW	0, 3.5, 10 and 80	Increased yield	Ozores and Stansly, 2005
Bell pepper	MSW	0, 30 and 60	No effect	Clark et al., 2000
Broccoli	DM	0, 10, 20 and 40	Increased yield	Roe and Gerald, 2010
Cucumber	DM	0 and 11	Increased yield	Nair and Ngouajio, 2010
Lettuce	SWM/SD/soya bean/peanut and crop residue	6.2, 12.5, 18.7 and 25	Increased yield	Chang et al., 2009
Muskmelon	DM	0, 10, 20 and 40	Increased yield	Roe and Gerald, 2010
Potato	DM	9 and 17.8	No effect	Barmaki et al., 2008
Potato	MSW	10, 20 and 29	Increase yield	Mkhabela and Warman, 2005
Squash	BS/MSW/YW	0, 7.1, 10.7 and 21.4	Increased yield	Ozores et al., 1994b
Tomato	Stover/cassava peels/PM	0 and 2.2	Increased yield	Taiwo et al., 2007
Tomato	SFC	0 and 83.9	Increased yield	Stoffella and Greatz, 2000
Tomato	SFC	0, 83.9 and 100	Increased yield	Stoffella and Greatz, 2000
Tomato	SM/DM/PM/MSW	0 and 9	Increased yield	Mehdizadeh et al., 2013
Pak choi	SWM/SD/soya bean/peanut and crop residue	6.2, 12.5, 18.7 and 25	Increased yield	Chang et al., 2009
Watermelon	DM/YW/BS/MSW	0, 3.3, 10 and 80	No effect	Ozores and Stansly, 2005

BS = biosolids; DM = dairy manure; FW = food waste; MSW = municipal solid waste; PM = poultry manure; SD = sawdust; SFC = sugar filter cake; SM = sheep manure; SWM = swine manure; YW = yard waste.

N fertilizer should be applied or planting should be delayed for 6–10 weeks to allow the compost to stabilize *in situ* (Obreza and Reeder, 1994). Crop injury has been linked to the use of poor-quality compost, such as that from early stages of the composting process (Ozores-Hampton et al., 1998a,b). The type and degree of plant injury are related to compost maturity or stability (Table 6.1). Maturity is the degree to which it is free of phytotoxic substances that can cause delayed seed germination or seedling and plant death; stability is the degree to which compost consumes N and O_2 in significant quantities to support biological activity and generates heat, carbon dioxide (CO_2) and water vapor that can cause plant stunting and yellowing of leaves. Plant stunting has often been attributed to high C:N ratio of the organic material before humification and plant injury from exposure to phytotoxic compounds such as volatile fatty acids and ammonia. Phytotoxin identification in compost extracts from fresh and 5-month-old MSW material showed that fresh compost contained acetic, propionic, isobutyric, butyric and isovaleric acids in the largest concentrations. Acetic acid at 300 ppm concentration inhibited the growth of cress seed (Table 6.1) (Ozores-Hampton et al., 1998a b).

Some vegetables such tomatoes are crops that are sensitive to high EC or soluble salts, especially when they are direct seeded. We recommend measuring the EC concentration of a saturation extract. If the EC is below 6.0 dS/m, no salt toxicity should occur. If the EC is above 6.0 dS/m, the amended soil should be leached with water before planting seeds (only a few crops can tolerate this EC level). In the desert areas where rainfall is limited, EC accumulation due to compost high in EC and/or with high application rates needs to be considered annually. Some of the EC in animal manures is from the nutrients salts, such as K, Ca and Mg. However, a compost in which a large portion of the EC is from sodium is not desirable in crop production.

6.6 Steps to success using compost in vegetable production

1. Compost must pass applicable federal and state law such as USEPA regulation 40 CFR Part 503 for windrow composting of biosolids: temperatures of 131ºF for 15 days and turned five times will eliminate human and plant pathogen, nematodes and weed seeds.
2. Meet "horticultural specification" based on the crop requirement (Table 6.1). Compost should be stable and mature to avoid N "rob" and phytotoxic reactions to chemicals (acetic, propionic and butyric acids).
3. Compost is not considered a fertilizer; however, significant quantities of nutrients (particularly P and micronutrients) become bioavailable with time as compost decomposes in the soil. Amending soil

with compost provides a slow-release source of nutrients, whereas inorganic synthetic fertilizer is usually water-soluble and is immediately available to plants. Therefore, it is important to determine the nutrient content by a compost-certified laboratory. Total N, P and K apply by the compost should be deducted from the total fertilizer N, P and K and micronutrient annual application rate.

6.7   Avoiding problems using compost in vegetable production

1. Use of immature compost can cause detrimental effects on vegetable growth. We recommend assaying compost for the presence of phytotoxic compounds or high EC using phytotoxicity germination test and seedling growth responses (Table 6.1).
2. High C:N compost can result in N-immobilization or "rob." Have the compost analyzed for C:N ratio. If it is above 20:1, some N fertilizer applied to the crop may be immobilized due to N-immobilization, possibly causing plant N deficiency. When using compost with C:N ratios higher than 20:1, N fertilizer should be applied or planting should be delayed for 6–10 weeks to allow the compost to stabilize in the soil.
3. Lack of equipment to spread compost in vegetable fields is a concern. We encourage compost facilities to play an active role in developing spreading equipment.

References

Aggelides, S.M., and P.A. Londra. 2000. Effects of compost produced from town wastes and sewage sludge on the physical properties of a loamy and a clay soil. *Bioresource Technology* 71:253–259.

Aram, K., and A. Rangarajan. 2005. Compost for nitrogen fertility management of bell pepper in a drip-irrigated plasticulture system. *HortScience* 40:577–581.

Barmaki, M.F., R.S. Khoei, Z. Salmasi, M. Moghaddam, and G.N. Ganbalani. 2008. Effect of organic farming on yield and quality of potato tubers in Ardabil. *Journal of Food, Agriculture and Environment* 6:106–109.

Bhattacharyya, P., K. Chakrabarti, and A. Chakraborty. 2003. Effect of MSW compost on microbiological and biochemical soil quality indicators. *Compost Science and Utilization* 11:220–227.

Bryan, H.H. and C.J. Lance. 1991. Compost trials on vegetables and tropical crops. *BioCycle* 27(3):36–37.

Bryan, H.H., B. Schaffer, and J.H. Crane. 1995. Solid waste for improved water conservation and production of vegetables crops (tomatoes/watermelons). In *Florida Water Conservation/Compost Utilization Program. Final Report*, edited by W.H. Smith, 1–14. Immokalee, FL.

Bulluck, III., L.R., M. Brosius, G.K. Evanylo, and J.B. Ristaino. 2002. Organic and synthetic fertility amendments influence soil microbial, physical, and chemical properties on organic and conventional farms. *Applied Soil Ecology* 19:147–160.

Chang, E., R. Chung, and Y. Tsai. 2009. Effect of different application rates of organic fertilizer on soil enzyme activity and microbial population. *Journal of Soil Science and Plant Nutrition* 53:132–140.

Clark, G.A., C.D. Stanley, and D.N. Maynard. 2000. Municipal solid waste compost (MSWC) as a soil amendment in irrigated vegetable production. *Transactions of the ASAE* 43:847–853.

Gallaher, R.N. and R. McSorley. 1994a. Management of yard waste compost for soil amendment and corn yield. In *The Composting Council's Fifth Annual Conference Proceedings*, 28–29. Washington, DC. 16–18 November 1994.

Gallaher, R.N. and R. McSorley. 1994b. *Soil Water Conservation from Management of Yard Waste Compost in a Farmer's Corn Field*. Gainesville, FL: IFAS, Agronomy Research Report AY-94-02.

Ghaly, A.E. and F.N. Alkoaik. 2010. Effect of municipal solid waste compost on the growth and production of vegetable crops. *American Journal of Agricultural and Biological Sciences* 5:274–281.

Guerrero, C., I. Gómez, J.M. Solera, R. Moral, J.M. Beneyto, and M.T. Hernández. 2000. Effect of solid waste compost on microbiological and physical properties of a burnt forest soil in field experiments. *Biology and Fertility Soils* 32:410–414.

Hernández, T., C. Chocano, J.L. Moreno and C. García. 2014. Towards a more sustainable fertilization: Combined use of compost and inorganic fertilization for tomato cultivation. *Agriculture, Ecosystems and Environment* 196:178–184.

Jackson, T.L., W. Brinton, D.T. Handley and M. Hutton. 2013. Residual effects of compost applied to sweet corn over two crop seasons. *Journal of the NACAA* 6:1–7.

Kostewicz, S.R. 1993. Pole bean yield as influenced by composted yard waste soil amendments. *Proceedings of the Florida State Horticultural Society* 106:206–208.

Kostewicz, S.R. and N.E. Roe. 1991. Yard waste and poultry manure composts as amendments for vegetable production. *HortScience* 26:6.

Lalande, R. B. Gagnon, and R.R. Simard. 2003. Papermill biosolids and hog manure compost affect short-term biological activity and crop yield of a sandy soil. *Canadian Journal of Soil Science* 83:353–362.

Li, Y., E. Hanlon, G. O'Connor, J. Chen, and M. Silveira. 2010. Land application of compost and other wastes (by-products) in Florida: regulations, characteristics, benefits, and concerns. *HortTechnology* 20:41–51.

Loper, S., A.L. Shober, C. Wiese, G.C. Denny, and C.D. Stanley. 2010. Organic soil amendment and tillage affect soil quality and plant performance in simulated residential landscapes. *HortScience* 45:1522–1528.

Massri, M. and L. Labban. 2014. Comparison of different types of fertilizers on growth, yield and quality properties of watermelon (*Citrllus lanatus*). *Journal of Agricultural Science* 5:475–482.

Maynard, A. 1995. Cumulative effect of annual additions of MSW compost on the yield of field-grown tomatoes. *Compost Science and Utilization* 3:47–54.

Mehdizadeh, M., E.I. Darbandi, H. Naseri-Rad, and A. Tobeh. 2013. Growth and yield of tomato (*Lycopersicon esculentum* Mill.) as influenced by different organic fertilizers. *Journal of International Agronomy and Plant Production* 4:734–738.

Mkhabela, M.S. and P.R. Warman. 2005. The influence of municipal solid waste compost on yield, soil phosphorus availability and uptake by two vegetable crops grown in a Pugwash sandy loam soil in Nova Scotia. *Agriculture, Ecosystems and Environment* 106:57–67.

Montemurro, F., G. Convertini, D. Ferri, and M. Maiorana. 2005. MSW compost application on tomato crops in Mediterranean conditions: effects on agronomic performance and nitrogen utilization. *Compost Science & Utilization* 13:234–242.

Nair, A. and M. Ngouajio. 2010. Integrating row covers and soil amendments for organic cucumber production: implications on crop growth, yield, and microclimate. *HortScience* 45:566–574.

Obreza, T. 1995. Solid waste for improved water conservation and production of vegetable crops (tomatoes/watermelons). In *Florida Water Conservation/ Compost Utilization Program. Final Report*, edited by W.H. Smith, 15–32. Immokalee, FL.

Obreza, T.A. and R.K. Reeder. 1994. Municipal solid waste compost use in a tomato/watermelon successional cropping. *Soil and Crop Science Society of Florida, Proceedings* 53:13–19.

Ozores-Hampton, M. 2012. Developing a vegetable fertility program using organic amendments and inorganic fertilizers. *HortTechnology* 22:743–750.

Ozores-Hampton, M. 2017. Guidelines for assessing compost quality for safe and effective utilization in vegetable production. *HortTechnology* 27:150.

Ozores-Hampton, M., F. Di Gioia, S. Shinjiro, E. Simonne and K. Morgan. 2015. Effects of nitrogen rates on nitrogen, phosphorous and potassium partitioning, accumulation and use efficiency in seepage-irrigated fresh market tomatoes. *HortScience* 50:1636–1643.

Ozores-Hampton, M. and H.H. Bryan. 1993a. Effect of amending soil with municipal solid waste (MSW) compost on yield of bell peppers and eggplant (Abstract). *HortScience* 28:463.

Ozores-Hampton, M. and H.H. Bryan. 1993b. Municipal solid waste (MSW) soil amendments: Influence on growth and yield of snap beans. *Proceedings of the Florida State Horticultural Society* 106:208–210.

Ozores-Hampton, M., H.H. Bryan and R. McMillan. 1994a. Suppressing disease in field crops. *BioCycle* 35(7):60–61.

Ozores-Hampton, M., T.A. Obreza, and G. Hochmuth. 1998a. Using composted wastes on Florida vegetables crops. *HortTechnology* 8:130–137.

Ozores-Hampton, M., T.A. Obreza, and P.J. Stoffella. 1998b. Immature compost used for biological weed control. *Citrus and Vegetable Magazine*. March 12–14.

Ozores-Hampton, M. and D.R.A. Peach. 2002. Biosolids in vegetable production systems. *HortTechnology* 12:356–340.

Ozores-Hampton, M., and P. Roberts, and P.A. Stansly. 2012. Organic pepper production. In *Peppers: Botany, Production and Uses*, edited by V. Russo, 165–175. Oxfordshire, UK: CABI.

Ozores-Hampton, M., B. Schaffer, and H.H. Bryan. 1994b. Nutrient concentrations, growth, and yield of tomato and squash in municipal solid-waste-amended soil. *HortScience* 29:785–788.

Ozores-Hampton, M. and P.A. Stansly. 2005. Effect of long-term organic amendments and soil solarization on pepper and watermelon growth, yield and soil fertility. *HortScience* 40: 80–85.

Ozores-Hampton, M.P., P.A. Stansly, and T.P. Salame. 2011. Soil chemical, biological and physical properties of a sandy soil subjected to long-term organic amendments. *Journal of Sustainable Agriculture* 353:243–259.

Pane, C., G. Celano, A. Piccolo, D. Villecco, R. Spaccini, A. Palese, and M. Zaccardell. 2015. Effects of on-farm composted tomato residues on soil biological activity and yields in a tomato cropping system. *Chemical and Biological Technologies in Agriculture* 2:1–13.

Roe, N.E. and C.C. Gerald. 2010. Effects of dairy lot scrapings and composted dairy manure on growth, yield, and profit potential of double-cropped vegetables. *Compost Science and Utilization* 8:320–237.

Roe, N.E., S.R. Kostewicz. 1992. Germination and early growth of vegetable seed in compost. In *Proceedings of National Symposium Stand Establishment of Horticultural Crops*, edited by C.S. Vavrina, 191–201.

Roe, N.E., S.R. Kostewicz and H.H. Bryan. 1990. Effects of municipal solid waste compost and fertilizer rates on broccoli (abstract). *HortScience* 25:1066.

Roe, N.E., P.J. Stoffella, and H.H. Bryan. 1993a. Utilization of MSW compost and other organic mulches on commercial vegetable crops. *Compost Science and Utilization* 1(3):73–84.

Rosen, C.J. and P.M. Bierman. 2016. Using manure and compost as nutrient sources for fruit and vegetable crops. *The University of Minnesota extension, Minneapolis.* 29 May 2020. https://extension.umn.edu/how/how-manage-soil-and-nutrients-home-gardens.

Salazar, I., D. Millar, V. Lara, M. Nuñez, M. Parada, M. Alvear, and J. Baraona. 2012. Effects of the application of biosolids on some chemical, biological and physical properties in an Andisol from southern Chile. *Journal of Soil Science and Plant Nutrition* 12:441–450.

Setyowati, N., Z., Muktamar, B. Suriyanti, and M. Simarmata. 2014. Growth and yield of chili pepper as affected by weed based organic compost and nitrogen fertilizer. *Journal on Advanced Science, Engineering and Information Technology* 4:84–87.

Shober, A. and G. Vallad. 2012. *Evaluating Compost and Lime Effects on Soil Organic Matter, Soil Microbial Communities and the Control of Fusarium Wilt in Commercial Tomato Grown in Florida's Sandy Soils.* Wimauma, USA.

Stentiford E. and M.A. Sanchez-Monedero. 2016. Past, present and future of composting research. *Acta Horticulturae* 1146:1–10. 3 May 2020 http://dx.doi.org/10.17660/ActaHortic.2016.1146.1.

Stoffella, P. 1995. Growth of vegetables. In *Compost Test Program for the Palm Beach Solid Waste Authority Project. Final Report*, edited by W.H. Smith, 2–34. Palm Beach, FL.

Stoffella, P.J. and D.A. Graetz. 2000. Utilization of sugarcane compost as a soil amendment in a tomato production system. *Compost Science and Utilization* 8:210–214.

Stoffella, P.J., Z.L. He, S.B. Wilson, M. Ozores-Hampton, and N.E. Roe. 2014. *Utilization of Composted Organic Wastes in Vegetable Production Systems.* Food & Fertilizer Technology Center, Technical Bulletins. Taipei, Taiwan. 3 May 2020. http://www.agnet.org/htmlarea_file/library/20110808105418/tb147.pdf.

Taiwo, L.B., J.A. Adediran, and O.A. Sonubi. 2007. Yield and quality of tomato grown with organic and synthetic fertilizers. *Journal of International Vegetable Science* 13:5–19.

Test Methods for Examination of Composting and Compost (TMECC). 2002. *U.S. Composting Council*, Bethesda, MD.

U. S. Department of Agriculture (USDA). 2019. Vegetables 2018 summary. National Agricultural Statistics Service Office of Adjudication, Washington, DC. April 23, 2020. https://downloads.usda.library.cornell.edu/usda-esmis/files/02870v86p/gm80j322z/5138jn50j/vegean19.pdf.

U.S. Department of Agriculture/Agricultural Marketing Service. 2020. National Organic Program. 11 Apr. 2020. https://www.ams.usda.gov/about-ams/programs-offices/national-organic-program

U. S. Environmental Protection Agency (USEPA). 1994. *A Plain English Guide to the EPA Part 503 Biosolids Rule*. EPA832-R-93-003. Sept. Washington, DC.

U. S. Environmental Protection Agency (USEPA). 1995. *A Guide to the Biosolids Risk Assessments for the EPA Part 503 Rule*. EPA832-B-93-005. Sept. Washington, DC.

U. S. Environmental Protection Agency (USEPA). 1999. *Biosolids Generation, Use, and Disposal in the United States*. EPA503-R-99-009. Sept. Washington, DC.

Zhang, M., Y.C. Li, and P.J. Stoffella. 2003. Nutrient availability in a tomato production system amended with compost. *Proceedings of 6th IS on Protected Cult.* Edited by G. La Malfa et al. *Acta Horticulturae* 614:787–797, ISHS.

chapter 7

Compost utilization in ornamental and nursery crop production

Craig S. Coker

Contents

Ornamentals are plants that are grown for decorative purposes in gardens and landscape design projects, as houseplants, cut flowers and specimen display. The cultivation of ornamental plants is called floriculture, which forms a major branch of horticulture. They are grown in containers (in greenhouses or out in open), in field plantings in nurseries (products are usually sold as "balled-and-burlapped"[1]) and in orchards.

Compost additions to soil bring biological, chemical and physical benefits. Some of the effects on soils are shown in Table 7.1 (McConnell et al., 1993). The organic matter added in the compost adds to soil organic carbon (SOC). Briefly, SOC acts as a binding agent in the formation of soil aggregates, and soil aggregate stability is important in maintaining soil structure (Bronick and Lal, 2005). Soil structure is the arrangement of parts within soil, the formation of soil granules and the pore space between them. The quantity, density and integrity of porous spaces between soils impact the soils' ability to support plant growth, as porosity improves air, water and root travel through the soil.

Improved water holding capacity is important to high-quality plant health. Studies report indicate that for every 1% of organic matter content, the soil can hold 16,500 gallons of plant-available water per acre of soil down to 1 feet deep (Hudson, 1994). Compost can also influence soil pH. The effect varies depending on the acidity or alkalinity of the soil itself, as well as due to the pH of the compost. While most compost tends to

Table 7.1 Compost effects on soil properties

Parameter	Rate (cy/1,000 sf)	Effect (% increase)
Organic matter	1.0–6.5	6–163
Water holding capacity	0.5–6.5	5–143
Bulk density	1.0–6.5	4–71
pH	1.0–6.5	0.8–1.4

have pH levels in the 6.0–7.0 range, some composts will have pH values closer to 8.0. In one study, adding 1.8 ounces of compost to 2.2 lb of soil, the pH increased from 5.2 to 5.4 in a Typic Xerochrept soil, while in a Typic Rhodoxeralf soil, it decreased from 7.3 to 6.7 (Filcheva, 2007).

7.1 Containerized floriculture

Many ornamental plants are raised in various containers ranging in size from 3.5 inches square to 20 gallons. Compared with roots of plants growing in the field, those growing in containers are exposed to a more stressful environment; for example, during active growth, a plant can extract all the available water in a typical container in as little as 1 or 2 days (Cabrera and Johnson, 2014a). Following irrigation, the medium is saturated at the bottom of the container. Roots located in the saturated medium exist without air until the plant uses enough water to create air spaces. As the medium dries, the soluble salt (electrical conductivity) concentration in the soil solution can increase to high levels. Key nutrients such as nitrogen (N) and potassium (K) are lost to both plant uptake and leaching and may become rapidly depleted if they are not supplied periodically. Growing medium temperatures can exceed 120°F and can fluctuate by 30°F or more between day and night.

A growing medium that optimizes the production of high-quality container plants is critical. In contrast to field production, the volume of medium available per plant is limited and must have acceptable physical and chemical characteristics that, when coupled with an intelligent management program, support plant optimum growth (Cabrera and Johnson, 2014a). Physical properties are often considered more important to a container medium because they cannot be easily changed after ingredients have been placed in a container. Chemical properties, however, can be altered. Table 7.2 summarizes characteristics of an acceptable good container medium.

Comparing Tables 7.1 and 7.2, one can see how compost-based container media can meet most of the characteristics of an adequate container medium. As compost is relatively heavy (i.e., has a high bulk density), compost-based container media often incorporate materials that reduce

Table 7.2 Characteristics of an adequate container medium to optimize plant growth

Adequate water holding capacity	Physical property
Adequate aeration and drainage	Physical property
High permeability to water and air	Physical property
Light weight	Physical property
Adequate fertility	Chemical properties
Pasteurized	Management practice
Inexpensive	Costs based on availability and transportation

the overall weight of the medium. These additives include perlite (a gray-white silicaceous volcanic rock, crushed and heated at 1,400°), vermiculite (a micaceous mineral, expanded by heating to 1,400°.), wood chips, bark or peanut hulls.

Organic matter (and thus, SOC) can be added using several different sources, including manures, peat moss or topsoil. Peat moss has been the preferred organic matter source in container media for decades, but it is not a renewable resource. In one study, growth of 24 perennial ornamental plants was evaluated using a commercially available peat-based soilless medium, amended with 25%, 50% or 75% compost generated from bio-solids and yard trimmings (Wilson and Mecca, 2004). Use of 100% compost in the medium increased plant growth of 11 species, reduced plant growth of 6 species and did not affect plant growth of the remaining 6 species tested when compared to the peat-based commercial control mix. Use of 50% compost in the medium increased plant growth of 8 species, reduced plant growth of 2 species and did not affect plant growth of the remaining 14 species tested when compared to the peat-based commercial control mix. Regardless of the plant species tested, compost amendments did not affect flowering or visual quality, and plants were still considered marketable. The study's results suggested that compost can be a viable alternative to peat as a substrate for containerized perennial ornamental production.

Each organic matter source has its individual characteristics, as shown in Table 7.3, which qualitatively compares several organic matter sources (compost, animal manure, peat and topsoil) on the basis of physical and chemical parameters, such as the content of nutrients, soluble salts and organic matter, and with regard to pH, bulk density and water-holding capacity (Evanylo, 2019).

Composts selected for use in a container medium should have the characteristics shown in Table 7.4 (United State Composting Council (USCC), 1996). Research and field experience have shown that the typical factors limiting the inclusion rate of compost in container media are exceedances

Table 7.3 Characteristics of different organic matter sources

Parameter	Compost	Manure	Peat	Topsoil
Nutrients	Medium–high	High	Very low	Low–medium
Soluble salts	Medium–high	Medium–high	Very low	Low
pH	Medium	Medium–high	Low–very low	Low–medium
Bulk density	Medium	High	Low	High
Water hold capacity	Medium	Low–medium	High–very high	Low
Organic matter	Medium–high	Medium–high	High–very high	Low

Table 7.4 Preferred compost characteristics for container media

Parameter	Value range
pH	5.5–8.0
Moisture content	35–55%
Particle size	Pass through 0.5 inches screen or smaller
Stability	Stable to highly stable to minimize shrinkage
Maturity	Very mature as measured with bioassay
Soluble salt concentration	Less than 3 dS/m (mmhos/cm)

of the values shown in Table 7.3. Compost made from yard trimmings, the organic fraction of municipal solid wastes (mostly food scraps) and biosolids have all been used to make container media. Solid waste composts have been used successfully at inclusion rates of 20–50%, biosolid composts at rates of 20–33% and yard trimming composts at rates of 33–47% (USCC, 1996).

Proper management of macronutrients, as well as micronutrients, is essential for the successful production of containerized ornamentals. The primary macro-elements N, phosphorus (P) and K are required in the highest amounts, and thus, of greatest importance in a fertilization program. In woody ornamentals, the elemental ranges of these primary nutrients can be used as a guideline: 2.0–2.5% N, 0.2–0.4% P and 1.5–2.0% K in uppermost mature leaves (Zinati, 2005).

Nitrogen is required in the highest amount of all the mineral elements. Plants may uptake N as either nitrate (NO_3^-) or ammonium (NH_4^+). Although NH_4^+ may be uptaken directly by the plant, it is also converted to NO_3^- in the substrate container media by the oxidation of NH_4^+ to NO_3^- and the oxidation of nitrite to NO_3^- (Troeh and Thompson, 2005). Composts used in container media tend to have NH_4^+ content of approx. 0.076%, NO_3^+ content of 0.034% and total N content of 1.7%,[2] and so additional nitrogenous fertilization is usually needed. There is usually adequate P (approx. 0.62%) in composts to meet plant fertilizer requirements.

Potassium is required at higher concentrations than P but lower concentrations than N; composts provide about 0.9% K, and so additional K fertilization is usually required.

Soluble salts can be an issue with compost use in container media. The presence of excessive soluble salts is perhaps the most limiting factor in the production of container crops. Generally, soluble salt accumulations result from the use of poor-quality irrigation water, overfertilization or growing media with an inherently high salt content. Composts can have high soluble salt content depending on feedstock sources, with poultry litter and food scraps composts having more soluble salts than yard trimmings composts. Soluble salt contents (in container media) of 0–1.0 deciSiemens per meter (dS/m) are considered low, 1.0–2.0 dS/m are considered adequate and greater than 3.0 dS/m are considered high (Cabrera and Johnson, 2014b). Soluble salts levels in composts have been measured from 0.1 to over 60 dS/m.

A potential problematic constituent in some compost made from horse or dairy manures is persistent herbicides. The four main persistent herbicides of concern are clopyralid (sold as Confront, Stinger and other brands), picloram (sold as Tordon or Grozon), aminopyralid (sold as Milestone, Forefront and other brands) and aminocyclopyrachlor (sold as Imprelis, Streamline and other brand names). These herbicides are designed to be persistent and systemic in their action against broadleaf weeds. They are often used by hay farmers to control broadleaf weeds in hay fields, and when that hay is fed to ruminants, it passes through the animal and ends up in the manure. These herbicides have been shown to be phytotoxic at concentrations in composts of less than 1.0 part per billion (Coker, 2013). The expressed result in a plant to the exposure from persistent herbicides is significant leaf curl, so bioassays using large-leafed plants such as fava beans (*Vicia faba*) are recommended before incorporating manure composts into container media blends.

7.2   Field-grown floriculture

As landscaping activities often use larger trees and shrubs, some ornamentals are grown in the field at nurseries. Harvesting those plants results in a loss of viable topsoil over time. While nursery soils may be enhanced and improved over time through the growth and incorporation of cover crops, this practice requires nursery fields out of production for periods of time. Compost can be utilized after harvesting or before planting to keep fields in production. The preferred characteristics of compost to be used in field nursery production are shown in Table 7.5 (USCC, 1996).

Compost stability is defined as "completeness of decomposition" (of the original wastes), where compost maturity is defined as "completeness of composting" (Coker, 2019). Composts that have less stability can be

Table 7.5 Preferred compost characteristics for field production

Parameter	Value range
pH	5.5–8.0
Moisture content	35–55%
Particle size	Pass through 1-inch screen or smaller
Stability	Stable to highly stable. Less stable composts can be used if planting is delayed
Maturity	Mature as measured with bioassay
Soluble salt concentration	Less than 3 dS/m (mmhos/cm)

used in field nursery production if they are incorporated into the soil 1–2 months before planting the ornamentals.

Composts with higher levels of soluble salts can be used in field production as the soluble salts will eventually leach out of the soil profile, although lesser application rates should be used with salt-sensitive species like dogwoods, narrowleaf conifers and ericaceous plants like *Rhododendron*, *Camellia*, azalea, *Pieris*, summer-flowering heathers (*Calluna*) and Japanese maples (*Acer*).

Composts can be introduced into field nursery production by broadcasting 50–100 tons per acre of compost and incorporating it into the top 6 inches of the soil profile, by amending the planting hole backfill when field-planting a container plant (mixing the compost and the backfill on a 50%/50% volume basis) or with fields in production, by applying the compost in a narrow band along the rows of trees or shrubs (known as side-banding).

Broadcast or side-banded field application of composts should be done through calibrated spreaders. Calibrating a spreader requires reliable estimates of both the amount applied and area covered. There are two common calibration techniques. The swath or load-area method involves measuring both the amount of compost in a typical spreader load and the land area covered by applying one load of compost. The tarp or weight-area method involves weighing the compost spread over a small surface and computing the amount of compost applied per acre (Natural Resources Conservation Service (NRCS), 2007).

One of the benefits of compost incorporation into field nursery production is the disease suppressive benefits of compost. Researchers have found that compost-amended soils offer biological control (biocontrol) which can be defined as the use of living organisms to depress the population of a pest. Long-term biocontrol activity can be obtained in bark mixes and in mixes prepared with composts. Composted pine bark mixes (30% or higher pine bark by volume) may support control for a year or more, but it depends on how it was aged or composted, and mix temperature.

Mixtures of pine bark (65% or higher) and composted sewage sludge (5–12% by volume) or other composts such as composted manures (5–8% by volume), composted leaves or green trimming wastes or composted hardwood bark (typically blended in at volumetric ratios of 12–25%) may provide control for 18 months or longer. These mixes have proven particularly effective against *Pythium* and *Phytophthora* root rots and *Thielaviopsis*[3] black root rot in perennial and nursery crops, especially when inoculated with specific biocontrol agents capable of inducing systemic resistance in plants (Hoitink, 2008).

Researchers have found that compost effects on soil physical properties in field nurseries differed among composts and that the effect on shrub growth was species-specific (Gonzalez and Cooperband, 2002). In this study, they amended field soils with three composts to evaluate their effects on soil physical properties and shrub biomass production. The application was either duck manure-sawdust (DM), potato cull-sawdust-dairy manure (PC) or paper mill sludge-bark (PMB) composts to a Plano silt loam soil using two application methods: 1 inch of compost incorporated into the top 6 inches of soil (incorporated-only) or 1 inch of compost incorporated plus 1 inch of compost applied over the soil surface (mulched). Three shrub species, *Spirea japonicum* "Gumball," *Juniper chinensis* "Pfitzeriana" and *Berberis thunbergii* "Atropurpurea," were grown. Mulched treatments resulted in 15–21% higher total SOC than the incorporated-only and nonamended control treatments. Bulk density decreased with increasing soil carbon contents. Greater aggregate stability and the formation of larger aggregates were related to increase in SOC. Field moisture retention capacity tended to be higher in the incorporated treatments compared to the mulched and nonamended control treatments. Compost-amended treatments increased saturated hydraulic conductivity sevenfold over that of nonamended control. There were no compost effects on shrub biomass until the second year of growth. *Berberis* was the only species to respond significantly and positively to compost application. Mulched DM compost produced 39–42% greater total *Berberis* biomass than the other compost treatments and the nonamended control.

Notes

1. By definition, balled-and-burlapped plants are transplants sold to the consumer after having been planted, dug up and wrapped. "Balled" refers to the root ball (that is, soil plus roots), which has been dug up, while "burlapped" refers to the wrapping material traditionally used for transporting tree and shrub deliveries.
2. Unpublished data based on averages of 2,182 compost samples analyzed in 2000–2005.
3. *Thielaviopsis* is a small genus of fungi in the order Microascales.

References

Bernal, M.P., S.G. Sommer, D. Chadwick, C. Qing, L. Guoxue, and F. Michel. 2017. Current approaches and future trends in compost quality criteria for agronomic environments. *Advances in Agronomy* 44:172–173.

Bronick, C.J. and R. Lal. 2005. Soil structure and management: a review. *Geoderma* 124(1–2):3–22.

Cabrera, R.I. and J.R. Johnson. 2014a. *Fundamentals of Container Media Management.* Fact sheet FS812. Rutgers University Cooperative Extension, New Brunswick, NJ.

Cabrera, R.I. and J.R. Johnson. 2014b. *Monitoring and Managing Soluble Salts in Ornamental Plant Production.* Fact sheet FS848, Rutgers University Cooperative Extension.

Coker, C.S. 2013. Composters defend against persistent herbicides. *BioCycle* 54(8):21–24.

Coker, C.S. 2019. *U.S. Composting Council Compost Operator Training Course,* Raleigh, NC.

Evanylo, G.E. 2019. *Compost Use.* Richmond VA: Mid-Atlantic Better Compost School, August.

Filcheva, E.G. and C.D. Tsadilas. 2007. Influence of clinoptilolite and compost on soil properties. *Communications in Soil Science and Plant Analysis* 33(3–4):595–607.

Gonzalez, R.F. and L.R. Cooperband. 2002. Compost effects on soil physical properties and field nursery production. *Compost Science & Utilization* 10(3):226–237.

Hoitink, H.J. and D. Lewandowski. 2008. Biological control of diseases of containerized Plants. *BioCycle* 49(7)32.

Hudson, B.D. 1994. Soil organic matter and available water capacity. *Journal of Soil and Water Conservation* 49(2):189–194.

McConnell, D.B., A. Shiralipour, and W.H. Smith. 1993. Compost application improves soil properties. *BioCycle* 34(4):61–63.

Natural Resources Conservation Service. 2007. Calibrating manure spreader application rates. 18 April. 2020. https://www.nrcs.usda.gov/Internet/FSE_DOCUMENTS/stelprdb1167137.pdf.

Troeh, F.R. and L.M. Thompson. 2005. *Soils and Soil Fertility.* Blackwell Publishing, Ames, IA.

U.S. Composting Council. 1996. *Compost Use Guidelines – Nursery. Field Guide to Compost Use,* USDA-CSREES Grant No. 91-COOP-1-6519, 51–67, Alexandria, VA

Wilson, S.B. and L.K. Mecca. 2004. Evaluation of compost as a viable medium amendment for containerize perennial production. *Acta Horticulturae* 659:697–703.

Zinati, G. 2005. *Nutrients and Nutrient Management of Containerized Nursery Crops.* Rutgers University Cooperative Extension, Fact Sheet E303. 18 April 2020. https://njaes.rutgers.edu/pubs/publication.php?pid=E303.

chapter 8

Compost utilization in turfgrass and lawn management

David Hill

Contents

According to the Royal Botanic Gardens at Kew, the turfgrass family (Poaceae) includes approx. 12,000 species in 700 genera. Of these, 37 species plus 18 genera of interspecific hybrids have been considered a suitable growth habitat to have been developed as turfgrasses. Turfgrasses contribute to major functional, recreational and aesthetic benefits for the ecosystems (Beard and Green, 1994; Beard, 2000) (Table 8.1). Turfgrass is widely grown for golf courses, lawns, parks, athletic field, commercial grounds, cemeteries and highway right-of-ways.

The most recent estimate of U.S. land coverage suggests that 1.9% of the United States is covered by these turfgrass areas. Home lawns are the largest category of turfgrass use. Turfgrass is a major component of managed landscapes and provides many functional, recreational, commercial and aesthetic benefits, including support of biodiversity, controlling erosion, dust reduction, dissipating solar heat, providing safety for athletic fields and improving quality of life (Beard and Green, 1994; Beard, 2000). Supporting growth within the U.S. turfgrass industry is a strong demand for the residential and commercial developments in

Table 8.1 Benefits of turfgrass and the soil, water, air, plant and human ecosystems

Functional		Recreational	Well being
Soil erosion control	Heat dissipation	Low-cost surfaces for physical and mental health	Mental health
Dust prevention	Air filtration	Decreased injury risk	Social harmony
Aerial particulate mitigation	Glare reduction	Family lawn activities	Community pride
Safeguard areas	Noise abatement	Community recreation	Human productivity
Natural filtration	Roadside safety	Community sports	Increased property value
Flood mediation	Crime control	Spectator entertainment	Aesthetic beauty
Organic pollution mitigant	Nuisance animal reduction		Compliments landscape
Topsoil restoration	Wildlife habitat		
Bioremediation	Pollen/weed control		
Carbon sequestration			

Source: Beard, 2020.

urban areas (Morris, 2006). There exist about 50 million acres of managed turfgrass in the United States, having an approximate value of $40–$60 billion. The turfgrass industry includes sod farms, lawn care services, lawn and garden centers and lawn equipment manufactures. Putting this into perspective, the estimated 31 million acres of irrigated turfgrass makes this the largest irrigated crop in the United States (Morris, 2006).

An area of turfgrass management that has only recently entered elevated consideration by researchers is that of soil health management, relative to the optimization of turfgrass production. In its simplest terms, this means actively managing soils by using natural or synthetic fertilizers or a combination of the two, soil cultivation practices and irrigation techniques to improve the quality of the soil in which turfgrass and most other plants grow. By treating the soil as a living system, and doing everything possible to improve and maintain its health, it appears that the health of the plants growing on a healthy soil will likewise be enhanced (Beard and Green, 1994; Beard, 2000). Inclusion of quality compost into these soil and crop management practices provides numerous benefits far beyond optimizing any or even a combination of all the above practices versus non-inclusion (Evanylo, 2005). This chapter intends to demonstrate that increased compost utilization in lawn and turfgrass management can, in fact, mitigate many of the environmental concerns and improve the economics of the turfgrass industry.

8.1   The history of turfgrass

A lawn has been defined as a section of a land area covered with mowed turfgrass plants. The development of turfgrasses by modern civilizations was to enhance the quality of life. As civilization developed, the more widely turfgrasses have been incorporated into daily life (Anonymous, 2020a). The concept of humans surrounding themselves with low-growing turfgrass is a trait learned from our ancient ancestors. In Africa, centuries ago, the low turfgrasses of the spacious savannas allowed humans to spot dangerous situations or animals. In England and Europe in the medieval times, grass-filled spaces with no trees around castles enabled sentries to spot any encroaching enemies. The 12th century shows evidence of the advent of cultured turfgrass lawns (Anonymous, 2020a). These low-growing, perennial grasses were kept low by animals grazing or using hand scythes. Sports was the first that used turfgrass in the late 12th and early 13th centuries. Bowling greens, finely laid, close-mown and rolled stretches of turfgrass for playing the game of bowls, were first used in the mid- to late 1200s. They have been attributed as the predecessors of our modern, fine turfgrasses used on tennis courts, croquet courts, badminton, golf putting greens and other field sports requiring highly manicured turfgrass (Anonymous, 2020a). The cricket of those times was the first team sport played on turfgrass. The European Renaissance period of the 15th century brings turfgrass within the private ornamental lawns and within the public parks spaces. In the 16th century, elaborate formal gardens graced the land of the elite.

During the 1500s, the evolution of the team sport, such as soccer, was played on the turfgrass areas of public greens in England (Anonymous, 2020a). According to Kaiser University, as far back as the 13th century, the Dutch played a game where a leather ball was hit with the intention of reaching a target several hundred yards away. The winner would be the player who reached the target with the fewest shots. However, the Scottish sport, which was known as "Golf," had one distinction that separates it from similar sports in history: the hole. When we talk about the modern game with 18 holes, golf history traces its origins back to 15th-century Scotland (Weathersby, 2017). Early immigrants to North America brought this lawn-related culture with them. They also brought along the seed to turfgrass that had been propagated in multiple areas.

Frederick Law Olmstead is commonly credited with sparking the migration of turfgrass lawns to the homes of ordinary North Americans. Not only did he design incredible public spaces, including New York's Central Park, but also a brought in a new aspect to residential green space. His design for the Chicago suburb of Riverside included a lawn for each home. Irrigation, in both in-ground automated systems and hose-end sprinklers, brought more efficient and effective water control to both commercial and private lawns. Currently, the focus on environment-friendly lawn care, with Integrated Pest Management (IPM) and Best Management Practices (BMP), added a new dimension to lawn maintenance. Advances in all these areas of turfgrass science continue at a rapid pace (Beard and Green, 1989).

8.2   What is compost?

Compost is defined as a product resulting from the controlled biological decomposition of organic material that has been sanitized through the generation of heat and stabilized to the point that is beneficial to plant growth (Vaughan et al., 2017). Compost should have no resemblance in physical appearance to the raw feedstocks from which it originates (U.S. Composting Council (USCC), 2001). The finished material is primarily organic matter (OM) that has the unique ability to improve the chemical, physical and biological characteristics of soils or growing media. It contains plant nutrients but is typically not characterized as a fertilizer (USCC, 2001). Compost is produced in many forms such as screened compost but also as compost tea, pellets and many other mixes and by-products.

8.3   The composting process

Compost is produced through the activity of aerobic (oxygen-requiring) microorganisms (USCC, 1996). For aerobic composting to occur, a minimum of 5% oxygen to ambient oxygen levels (~20%) must be present within the center of the pile or windrow. The microbes necessary for decomposition require the proper combination of oxygen (5%–20%), moisture (35%–55%) and food (C:N ratio of 25 to 30 parts carbon to every 1 part nitrogen (N)) to thrive and multiply. When these variables are maintained at optimal levels, the biological compost decomposition process is significantly accelerated. The microorganisms generate heat, water vapor and carbon dioxide as they transform raw materials into a stable soil conditioner. Active composting is typically characterized by a high-temperature phase that sanitizes the product and allows a high rate of decomposition followed by a lower-temperature phase that allows the product to stabilize while still decomposing at a slower rate (USCC, 1996).

8.4 Compost feedstocks suitable for use with turfgrass

Compost can be produced from many "feedstocks" (the raw organic materials, such as leaves, grass, manures and/or food scraps). State and federal regulations exist to ensure that only safe and environmentally beneficial composts are marketed (USCC, 2001). Compost types and suitability for sod and turfgrass include the following, primary categories:

Yard sourced materials: feedstocks will include leaves, grass clippings, ground brush, tree limbs, Christmas trees and various types of other OM originating from homeowner or professionally managed sites. Typically, yard sourced feedstocks are lower in nutrient value; however, they can be higher in contamination, due to the variability of the source of the feedstocks.

Agricultural by-products: feedstocks include field residues, animal manures, animal bedding, left-overs from crop processing, as well as fruit and vegetable culls; composts produced from agricultural by-products, especially manures, tend to be higher in nutrient value than some other feedstocks; however, manures are also higher in electrical conductivity (EC; soluble salts). Compost from this source is useful in field applications or incorporated into the soil.

Biosolids: these feedstocks are the processed wastewater solids derived from publicly and privately operated wastewater treatment plants. Of the three feedstock sources, biosolids are the most highly regulated and the most highly tested. The U.S. Environmental Protection Agency (USEPA) CFR 40 Part 503 regulations specify rules and regulations regarding the production, testing and distribution parameters for biosolids compost (USEPA, 1994, 1995). Only compost meeting "Class A" is acceptable for general distribution. This means it can be distributed the same way as any compost or mulch, for that matter, and can be used for any labeled purpose. Biosolids compost has the highest nutrient content of the three types of compost. They also have higher Fe (iron) levels, which is necessary for chlorophyll production in plants and especially suited for turfgrass. The Fe in biosolids compost creates a greener, denser turfgrass without pushing fleshy N growth. Along with a higher nutrient content, higher soluble salt levels are often found in biosolids compost relative to the others, making them highly suitable for field application and soil incorporation.

Food residuals/food waste: the term "food residuals" pertains to food that was not used for its intended purpose. These include plate waste (i.e., food that has been served but not eaten), spoiled food or peels and rinds, which are generated from households or institutional, commercial and industrial and governmental (ICIG) entities (U. S. Environmental Protection Agency (USEPA), 2020).

Municipal solid waste (MSW): this feedstock primarily consists of residential and commercial waste that has not been separated at the source to segregate paper, glass, plastics, recyclables, etc. from the mix. MSW intended for composting are separated at the collection facility to remove these items. Due to the high concentration of paper content in this feedstock type, the nutrient content is typically low, the pH high and it often has a higher rate of contamination from inert materials. MSW compost is considered the lowest value and the cheapest compost available. It is generally most suitable for field application in agricultural use, silviculture or forest reclamation, landfill, mine or spoils mediation or other field applications, such as roadside vegetation management (Shiralipour, et al., 1992).

Industrial by-products (IBP): they are residual materials from industrial, commercial, mining or agricultural operations that are not a primary product and not produced separately in the process (Minnesota Pollution Control Agency, 2014). These include various residuals that result from manufacturing, industrial and institutional sources. They include biodegradable packaging material, wood, pharmaceutical waste and paper. By-products from farms, industry and cities have nutrient value for crop production, but prudent management practices are needed to protect environmental quality (Sullivan, 2007). The benefits and attributes of IBP composts are highly variable from source to another and are completely dependent on the source of the feedstock (USCC, 2001).

8.5   Compost benefits to soil health in turfgrass production

Turfgrass is an excellent indicator of soil health and is often considered the cornerstone of an adequate fertility program. According to Landschoot (1997), consider the use of compost as a soil amendment to increase the turfgrass performance. Good quality compost use in clay soils will improve structure, reduce surface crusting and compaction, promote drainage and provide nutrients. In sandy soils, compost increases water and nutrient retention, supplies additional nutrients and increases microbial activity. Desirable physical and chemicals characteristics for compost in turfgrass are shown in Table 8.2. Physical and chemical compost attributes promote faster turfgrass establishment, improved turf density and color, increased rooting and less need for fertilizer and irrigation. Soil microorganisms are also affected by soil pH. Despite most fungi having adapted to a wide range of pH values, the bacteria and actinomycetes function best when

Table 8.2 Physical and chemical characteristics for compost application in turfgrass[z]

Physical attributes	
Color	Brown to black
Size (surface applications)	0.25–0.35 inches
Size (incorporated)	0.25–0.5 inches
Odor	Earthy
Moisture content	30%–50%
Chemical properties	
Carbon:nitrogen ratio	Below or equal to 30:1
Nitrogen	0.5%–3.0%
Phosphorus	Greater than 0.2%
pH	6.0–8.0
Metals	Determined by state and federal agencies
Organic matter	Greater than 30%
Ash content	Less than 70%
Soluble salts	Depends on turfgrass species, salt type, concentration and application method. Consult test laboratories or other expert to determine salt affects

[z] Use this information only as a general guide. Some composts have properties that do not fall within these guidelines and still are acceptable in certain situations. However, many passed these criteria and still have serious drawbacks.
Source: Landschoot, 1997.

soils are moderately acid to slightly alkaline. While there are many beneficial fungi, most turfgrass diseases resulting from pathogens are caused by fungi. Additionally, soil bacteria play a significant role in decomposition of OM and N transformations such as nitrification and denitrification.

8.6 Benefits of compost in turfgrass growth and development

Some of the most desirable results of compost use include the improved color, vigor and durability of turfgrass (Anonymous, 2020b). This is only partially a result of increased nutrient but also the addition of non-nutrient chemistry, such as humic compounds, and beneficial microbes such as fungi (i.e., mycorrhizae), bacteria (i.e., actinomycetes) and OM.

The physical properties of compost may increase soil porosity, water holding capacity, tilth, friability and particle adhesions through microbial activity affecting soil properties and plant growth (Duong, 2013). By increasing the soil porosity, water and air infiltration is increased.

Compost in clay soils will improve soil structure, reduce surface crusting and compaction, promote drainage and provide nutrients. In sandy soils, compost increases water and nutrient retention, supplies nutrients and increases microbial activity. These improvements promote faster turfgrass establishment, improve turfgrass density and color, increase rooting and reduce requirements of fertilizer and irrigation (Landschoot, 1997). When the soil OM (SOM) content is low, the soil is unproductive, and crops, trees or turfgrass have limited energy sources required to promote plant growth. When EC, chlorides and other nutrients are very high, the microbial biomass in the soil promoting aeration and decomposition of dead plant material (turning into SOM) will not survive, thereby significantly increasing the potential for soil compaction. When the soil is compacted, neither sufficient oxygen (air) nor water or nutrients can reach the root zone, thus the plant growth slows, and the plant may ultimately die. The desired goal of 5% SOM in the root zone can be achieved using compost, either in bed preparation prior to planting or in topdressing after turfgrass has been established.

Chemical properties of compost may modify or stabilize pH of the soil. The addition of compost to soil may adjust the pH of the final soil blend. However, depending on the pH of the compost and of the native soil, compost addition may raise or lower the pH in soil/compost blends. Therefore, the addition of a neutral to slightly alkaline compost to an acidic soil will increase soil pH if added in appropriate quantities. When the pH is outside from the optimal range, either too high or too low, the turfgrass will show symptoms such as yellowing, stunting, and in extreme cases, dying off. These are symptoms of nutrient deficiency or even toxicity, due to the plants' inability to uptake requirement and/or allowing them to concentrate to toxic levels. Consequently, turfgrass soils should be maintained in a slightly acid to neutral condition such as pH 5.9–7.0 (Anonymous, 2020c). In specific conditions, compost has been found to affect soil pH even when applied at quantities as low as 10–20 tons per acre (USCC, 2001). The incorporation of compost also can buffer or stabilize soil pH, whereby it will more effectively resist the pH change. Increase of cation exchange capacity (CEC) in compost will improve the CEC of soils, enabling it to retain nutrients longer. It will also allow turfgrass to utilize nutrients more effectively, while reducing nutrient loss by leaching. For this reason, the fertility of soils is often tied to their OM content. Improving the CEC of sandy soils by adding compost can greatly improve the retention of plant nutrients in the root zone. Compost contains a considerable variety of macro and micronutrients. Normally, it can provide sources of N, phosphorous (P) and potassium (K). Compost also contains micronutrients such as calcium (Ca), magnesium (Mg) and sulfur (S). These are commonly known as the major micronutrients and are essential for plant growth. Since compost contains relatively stable sources of OM, these nutrients are supplied in a slow-release, bio-converted form. On a pound-by-pound basis, large quantities of nutrients are not typically found in compost in comparison to most commercial fertilizers. However, compost is usually applied at significantly greater rates; therefore, it can have a significant cumulative effect on nutrient availability. The addition of compost can affect both fertilizer and pH adjustment (lime/sulfur addition). Compost not only provides some nutrition, but often makes current fertilizer programs more effective. Compost also adds beneficial chemical compounds resulting from the biological decomposition of volatile organic acids, converting them to humic compounds and nitrate-N, which provide significant benefits to the plants and biome within the soil (USDA, 2000). Humic substances (HS), especially humic acid and fulvic acid, are the result of microbial organic decomposition of plant and microbial remains. Humic acid is defined as a series of highly acidic, relatively high-molecular-weight, yellow to black colored substances occurring ubiquitously in nature (Dou et al., 2020). Humic acid benefits plants by improving soil structure and fertility and has been the major player in HS. Fulvic acids are made from are all elemental ions. It is water soluble in any pH and offers ionic trace minerals that are both electrically charged, thereby attracting nutrients, and small enough to pass through any cell wall easily (Matyukhin and Ironclad, 2020). These substances make up approximately 60% of SOM (Handler, 2020). Humic acids are extremely important as a medium for transporting nutrients from the soil to the plant because they can hold onto ionized nutrients, preventing them from soil leaching. Humic acids are also attracted to the depletion zone of the plant root zone. When humic acids enter the roots systems, they bring the water and nutrients that the plants require. Through chelation, HS increase the availability of cations to plants (Meléndrez, 2009). There are numerous claims about how humic acid applications benefit turfgrass and soil. The most consistently observed effect from humic acid applications, however, is a stimulatory effect, usually measured in root and shoot plant mass increases (Hoiberg, 2018). Topdressing turfgrass with a good quality, finished compost, aside from the aforementioned benefits, adds humic and fulvic acids to the soil.

"The enemy of my enemy is my friend." This ancient platitude holds true for competing microbes in the rhizosphere of turfgrass. When pathogenic microbes and beneficial microbes compete for the same nutrient sources, pathogenic microbes will decline when conditions are favorable for the beneficial microbes. Microorganisms occur in turfgrass soils in extremely high populations, making soils among the most dynamic sites of biological activity in nature (Nelson, 2020). Studies have shown the potential for compost amendments to reduce the severity and incidence of a wide variety of turfgrass diseases when applied either as a topdressing, a winter cover, a root zone amendment or as an aqueous extract (Nilsson, 2005).

Microbial and macrofauna communities play pivotal roles in the functioning of plants by influencing the plant physiology and development (Mendes et al., 2013). The area of interaction between the soil and plant roots is known as the rhizosphere (Broeckling et al., 2019). The introduction of compost into the production and maintenance of turfgrass provides an enormous diversity of microorganisms into rhizosphere microbiome in the soil. Microorganisms range in size from the small one-celled bacteria, algae, fungi and protozoa to the more complex, beneficial nematodes and micro-arthropods to the macrofauna, such as earthworms, insects, small vertebrates and plants (Ingham, 1999). The rhizosphere is a microzone in the soil where the plants and microbes form a symbiotic relationship. Plant roots secrete compounds such as sugars and amino acids, and microbes ingest and breakdown these compounds. It has been stated that up to 44% of the compounds that plants make from photosynthesis can end up in the rhizosphere (George, 2019). As these microorganisms ingest food, grow, multiply and move through the soil, they can provide cleaner water, air, healthy plants and moderated water flow (Ingham, 1999). There are many ways that the soil food web is an integral part of landscape processes. Soil microorganisms decompose organic compounds, including animal manure, plant residue and pesticides, preventing them from entering water and becoming pollutants (Ingham, 1999). Microorganisms sequester N and other nutrients that might otherwise enter groundwater and fix N from the atmosphere, making it available to plants. Many microorganisms enhance soil aggregation and porosity, thus increasing infiltration and reducing runoff (Ingham, 1999). Excess fertilizer use has been demonstrated to cause an increase in pest and pathogen levels. As a corollary, fertilizer reduction decreases disease pressure (Magazzi, 2013).

8.7 Compost considerations for use on turfgrass production

Compost has a broad range of valuable characteristics. Therefore, the benefits and versatility of compost can be realized in a wide variety of uses (USCC, 2001). Generally, turfgrass is susceptible to pests and disease (Landschoot, 1997). Since athletic turfgrass is heavily used and vigorously managed, it may become more susceptible to mechanical, pest and disease injury. Most commercial composts are of high quality; however, compost intended for special use should have a higher level of processing and degree of testing. Using compost as a soil amendment is an embraced practice for improving turfgrass performance in marginal or poor soils. Incorporating quality compost into heavy clay soils will improve soil structure, reduce surface crusting and compaction, promote drainage and provide nutrients. Sandy soils will benefit from compost by increasing water and nutrient retention, supplying nutrients and increasing microbial activity. The benefits from incorporating compost into these soils will include faster turfgrass establishment, improved turfgrass density and color, expanded root system and reduced fertilizer and irrigation requirements (Landschoot, 1997). Ideally, the best soil for turfgrass will have at least a 5% SOM content, and compost can be an optimal material to build SOM. Five percent SOM might take a significant amount of time to build in the soil; however, it will make a significant difference in the growth and vigor of the turfgrass.

Commercial composting facilities should consider adoption of the USEPA standards for the Process to Further Reduce Pathogens (PFRP) (USEPA, 1994, 1995). The USEPA standards indicate that compost must achieve an average temperature of 131° for the first 5 days in windrow composting process and should be turned five times and have the initial C:N ratio between 25:1 and 40:1. For an in-vessel or static aerated pile, compost should maintain a temperature of between 131°F and 170°F for 3 days. For compost intended to be listed for organic designation, accurate temperature records are required to satisfy the National Organic Program (NOP) standards (Rittenhouse, 2015).

8.8   Uses for compost on turfgrass production and maintenance

Compost use on turfgrass can be separated into two categories: turfgrass for lawns and turfgrass for athletic field use. Requirements for each are generally similar; however, considerations for each diverge along the lines of cost and functionality. Lawns address the primary functions of improved aesthetics, minimizing environmental impact. In sports turf, a quality playing surface will be appreciated by athletes and the appearance by spectators. A dense, wear-resistant turfgrass is essential to provide safety by cushioning falls and ensuring optimal footing. A primary contributor to a quality playing surface is the turfgrass. The turfgrass best suited for a given field depends largely on the demands imposed on the turfgrass, capital available for maintenance and the quality of surface desired (Miller, 2020). Requirements for athletic turfgrass are more demanding than those for lawn turfgrass. The sports/athletic uses are more physically demanding due to the need for faster turfgrass recovery and demand that fields be uniform to help reduce potential athletes' injury.

8.9   Compost use as turfgrass top dressing in lawn and athletic fields

Topdressings physically improve the soil structure, improving drainage and aeration, and may also be used to incorporate OM. Topdressing with compost is unique as a landscape maintenance practice because it closes the loop in the ecological cycle of sustainability. Compost topdressing utilizes compost (diverted and processed from the waste stream) and applies it to the soil as an amendment to improve the entire soil structure. Compost topdressing also reduces and enhances fertilizer and pesticide inputs and reduces soil compaction (Schuler, 2015). Topdressing turfgrass with compost accomplishes multiple benefits including the addition of macro and micronutrients, soil water retention, soil porosity, biochemical supplementation, nutrient augmentation, friability and nutrient uptake (Platt, 2016). Composts used for lawn turfgrass management can have different physical characteristics than those used on athletics turfgrass. Consequently, athletic turfgrass will require additional and specific testing. An example is shown in Table 8.3 for professional golf course use. To determine the volume of compost required for topdressing, use the following formula: area (A) to be covered, estimated in square feet (ft^2), multiplied by the desired depth (D) of compost to be applied or calculated as inches multiplied by 0.0031 = number of cubic yards (yd^3) required. Example, A = 10,000 ft^2 × 0.5 inches × 0.0031 = 15.5 yd^3 (Green Mountain Compost, 2020).

Typically, particle size is the major differentiating characteristic for composts produced for topdressing lawns versus compost used for topdressing athletic turfgrass. In residential/commercial turfgrass topdressing, compost is usually screened to a particle size of 1/4–3/8 inches in contrast to sports/athletic turfgrass, where turfgrass is mowed at shorter heights, particle sized will be screened to 1/8–1/4 inches.

When compost is being used in commercial professional golf courses, strict production and testing requirements have been established by the U.S. Golf Association (USGA) which is the U.S. national association of golf courses, clubs and facilities governing body of golf industry in the United States and Mexico. The most used organic component in commercial turfgrass for golf courses is a peat. There are a variety of other organic product sources that can be used in commercial turfgrass for golf courses such as rice hulls, finely ground bark, sawdust or other organic

Table 8.3 Compost volume requirements based on the depth of topdressing in turfgrass application

Depth (inches)	Amount needed (cubic yards/1,000 sq. ft.)
0.13	0.4
0.25	0.8
0.38	1.2
0.50	1.5
0.75	2.3

Source: Anonymous, 2020d.

waste that are acceptable if composted through a thermophilic stage to a mesophilic stabilization phase, and with the approval of the soil physical testing laboratory. Additionally, composts shall be aged for at least 1 year. Furthermore, the root zone mix with compost as the organic amendment must meet the physical properties as defined in these recommendations. Composts can vary not only with source, but also from batch to batch within a source. Extreme caution must be exercised when selecting a compost material to be utilized in commercial turfgrass for golf courses. Composts must be shown to be non-phytotoxic using a bentgrass (*Agrostis stolonifera* L.) or bermuda grass (*Cynodon dactylon* (L.) Pers.) bioassay on the compost extract. To avoid the future creation of alisol, a poorly draining soil with a subsurface of clay, there is a critical need for an accurate blend of soil components resulting in a homogenous root zone planting media. Therefore, all root zone mixes must be blended off-site, in advance, and all at one time. A quality control program during golf course construction is recommended. Arrangements should be made with a competent laboratory to routinely check gravel and/or root zone samples brought to the golf course construction site. Some tests can be performed on site with the proper equipment such as sand particle size distribution (Hummel, 1993).

8.10   Economic impact of compost use in turfgrass

The Massachusetts Development Complex (Devens, MA; www.devens community.com/recreation/fields-facilities) is the largest sports venue in New England. Comprised of 13 contiguous soccer fields and covering 44 acres, it was able to save $38,000 in operational costs over previous years by incorporating the use of compost in its turfgrass topdressing program (Hussey and Harrison, 2005). Prior to topdressing with compost in 2001, expenses for fertilizers and pesticides cost were $275,000/year. Water usage was 3 million gallons/year and the fields required reseeding three to four times/season. Three years later, in 2004, pesticide and fertilizer costs had decreased to $28,000/year and water use was lower by 83% (2.5 million gallons). Additionally, field reseeding requirements declined by 66%. On top of the cost reductions, the MD reported that "the fields have never looked better" (Hussey and Harrison, 2005). The above compost benefits as a topdressing in turfgrass are demonstrated in Table 8.4.

Turfgrass has been in existence for millennia and has been a significant part of civilized life for hundreds of years. It continues to grow in significance with an estimated 50,000 acres of turfgrass in the United States with a projected value of $40–$60 billion. Of the 12,000 species of the *Poaceae* grass family, there are 37 species plus interspecific hybrids that have been deemed to be suitable for turfgrass. The interrelationship between compost amended soil and that used as a topdressing creates a highly beneficial, dynamic response between the two. This results in

Table 8.4 Improved resource efficiency due to compost turfgrass topdressing applications

Inputs	2001	2004	Resource reduction
Seed and chemicals	$75,000	$28,000	$47,000
Water usage (gal)	3,000,000	500,000	2,500,000
Reseeding frequency (number/season)	3–4	1	2–3

a sturdier, more disease resistant and more aesthetically pleasing turf-grass with reduced costs overall to maintain. The use of compost results in enhanced physical soil structure, increased beneficial biological diversity and augmented chemical properties resulting from optimization of the other two properties.

This chapter has shown that, while there are many feedstocks, recipes and process methods for compost production, some are more suited to specific turfgrass species and or uses; any good quality compost will benefit managed turfgrass when used per recommendations versus non-use. Ultimately, evidence is shown that nourishing the soil is more effective, less costly and longer lasting than targeting nourishment of the plant. Additionally, the chapter shows specifically that the use of compost can significantly reduce water requirements by 75% and reduce chemical and reseeding costs by 63% while improving the look and durability of sports turfgrass.

References

Anonymous. 2020a. The history of the lawn. The Lawn Institute. 12 June 2020. https://www.thelawninstitute.org/pages/education/lawn-history/lawns-and-lawn-history/.

Anonymous. 2020b. Compost application to sports fields. SportsTurf Managers Association. June 3, 2020. https://www.stma.org/sites/stma/files/STMA_Bulletins/Compost_Final.pdf.

Anonymous. 2020c. Soil pH and liming. Umass Center for Agriculture, Food and Environment. June 12, 2020. https://ag.umass.edu/turf/fact-sheets/soil-ph-liming.

Anonymous. 2020d. Topdressing. Sports Field Management Cornell University. June 14, 2020. http://www.safesportsfields.cals.cornell.edu/topdressing.

Beard, B., and R.L. Green. 1994. The role of turfgrass in environmental protection and its benefits to humans. *Journal of Environmental Quality* 23(3):452–460.

Beard, J., and R.L. Green. 1989. Turfgrass and golf course benefits - A scientific assessment. *USGA Green Section Record*, 26–30. 12 June, 2020. http://pesticidetruths.com/wp-content/uploads/2018/05/Reference-Beard-Benefits-Of-Turfgrass-Turfgrass-Golf-Course-Benefits-A-Scientific-Assessment-USGA.pdf.

Beard, J.B. 2020. Field of dreams – The evolution of turfgrass sod. The Lawn Institute. 12 June 2020. https://www.thelawninstitute.org/pages/education/lawn-history/field-of-dreams/.

Broeckling, C.D., M.W. Paschke, J.M. Vivanco, and D. Manter. 2019. Rhizosphere ecology. *Encyclopedia of Ecology* 3:3030–3035. June 13, 2020. https://www.sciencedirect.com/science/article/pii/B9780080454054005401?via%3Dihub

Dou, S., J. Shan, X. Song, R. Cao, M. Wu, C. Li, and S. Guan. 2020. Are humic substances soil microbial residues or unique synthesized compounds? A perspective on their distinctiveness. *Pedosphere* 30(2):159–167. June 14, 2020. https://www.sciencedirect.com/science/article/abs/pii/S1002016020600017?via%3Dihub.

Duong, T. 2013. *Compost Effects on Soil Properties and Plant Growth.* PhD diss. University of Adelaide. June 14, 2020. https://digital.library.adelaide.edu.au/dspace/bitstream/2440/81916/8/02whole.pdf.

Evanylo, G.K. 2005. Impacts of compost, manure, and commercial fertilizer on soil and water quality, and crop production. June 13, 2020. https://portal.nifa.usda.gov/web/crisprojectpages/0181956-impacts-of-compost-manure-and-commercial-fertilizer-on-soil-and-water-quality-and-crop-production.html.

George, H. 2019. The benefits of using soil inoculants and microbes in the garden. *Gardener's Path.* June 13, 2020 https://gardenerspath.com/how-to/composting/benefits-soil-inoculants.

Green Mountain Compost Calculator Anonymous 2020. https://www.green-mountaincompost.com/compost-calculator/

Handler, J. 2020. Benefits of adding humic acid to soil for plants. *All Natural Ideas.* June 13, 2020. https://allnaturalideas.com/humic-acid/.

Hoiberg, A. 2018. Benefits to turfgrass from humic acid applications. *Calcium Products.* June 14, 2020. https://www.calciumproducts.com/benefits-to-turfgrass-from-humic-acid-applications/.

Hummel, Jr., N. 1993. USGA recommendations for putting green construction. *USGA Green Section Records.* June 13, 2020. https://gsrpdf.lib.msu.edu/ticpdf.py?file=/1990s/1993/930301.pdf.

Hussey, M. and C. Harrison. 2005. Compost for turfgrass: multifaceted organic ally. *Sportsturf.* June 13, 2020. https://sturf.lib.msu.edu/page/2005aug11-20.pdf.

Ingham, E.R. 1999. The soil biology primer. Chapter 1. The soil foodweb. NRCS Soil Quality Institute, USDA. June 13, 2020 https://www.nrcs.usda.gov/wps/portal/nrcs/detailfull/soils/health/biology/?cid=nrcs142p2_053868.

Landschoot, P. 1997. Using composts to improve turf performance. June 12, 2020. https://extension.psu.edu/using-composts-to-improve-turf-performance.

Magazzi, J. 2013. Soil health: the next big trend. *Turf Magazine.* June 13, 2020 https://www.turfmagazine.com/services/soil-health-the-next-big-trend/.

Matyukhin, V. and P. Ironclad. 2020. Fulvic acid benefits and one weird thing. June 13, 2020 https://www.purehimalayanshilajit.com/fulvic-acid/.

Meléndrez, M. 2009. Humic acid: The science of humus and how it benefits soil. *Eco Farming Daily.* June 13, 2020. https://www.ecofarmingdaily.com/build-soil/humus/humic-acid/.

Mendes, R., P. Garbeva, and J.M. Raaijmakers. 2013. The rhizosphere microbiome: significance of plant beneficial, plant pathogenic, and human pathogenic microorganisms. *FEMS Microbiology Reviews* 37(5):634–663 June 13, 2020. https://academic.oup.com/femsre/article/37/5/634/540803.

Miller, M. 2020. Tough turf. *Grounds Maintenance.* June 13, 2020. http://www.grounds-mag.com/mag/grounds_maintenance_tough_turf/.

Minnesota Pollution Control Agency. 2014. Guidelines for managing industrial by-products from food, beverage and agro-industrial processing facilities. June 12, 2020. https://www.pca.state.mn.us/sites/default/files/wq-lndapp2-03.pdf.

Morris, K.N. 2006. The national turfgrass research initiative. *USGA Green Section Record*, 26–30. 12 June, 2020. http://gsrpdf.lib.msu.edu/ticpdf.py?file=/2000s/2006/060926.pdf.

Nelson, E. 2020. Microbiology of turfgrass soils. *Ground Maintenance*. June 13, 2020. http://grounds-mag.com/mag/grounds_maintenance_microbiology_turfgrass_soils/.

Nilsson, J. 2005. Using biological methods in the golf industry. *BioCycle* 46(6):28. June 13, 2020. https://www.biocycle.net/2005/06/15/using-biological-methods-in-the-golf-industry/.

Platt, B. 2016. The benefits of composting & compost use. Institute for Local Self-Reliance. June 13, 2020. https://ilsr.org/benefits-composting-compost/.

Rittenhouse, T. 2015. Tipsheet: compost. *ATTRA Sustainable Agriculture*. June 13, 2020. https://www.ams.usda.gov/sites/default/files/media/Compost_FINAL.pdf.

Schuler, K. 2015. Compost topdressing & sustainable lawn care. *Turf Magazine*. June 13, 2020. https://www.turfmagazine.com/general-turf-care/compost-topdressing-sustainable-lawn-care/.

Shiralipour, A., D.B. McConnell, and W. Smith. 1992. Uses and benefits of MSW compost: a review and an assessment. *Biomass and Bioenergy* 3(3–4):267–279.

Sullivan, D.M. 2007. Utilization of municipal and industrial byproducts in agriculture. U. S. Department of Agriculture Research, Education and Economics Information Systems. June 12, 2020. https://portal.nifa.usda.gov/web/crisprojectpages/0174902-utilization-of-municipal-and-industrial-byproducts-in-agriculture.html.

U.S. Composting Council. 1996. Field guide to compost use. 23 Apr. 2020. http://compostingcouncil.org/admin/wp-content/plugins/wp-pdfupload/pdf/1330/Field_Guide_to_Compost_Use.pdf.

U.S. Composting Council (USCC). 2001. USCC factsheet: compost and its benefits. *WasteExchange.Org*. 12 June 2020. http://www.wastexchange.org/upload_publications/BenefitsofCompost.pdf.

U.S. Department of Agriculture (USDA). 2000. Part 637 environmental engineering national engineering handbook. *Natural Resources Conservation Services*. June 14, 2020 https://www.wcc.nrcs.usda.gov/ftpref/wntsc/AWM/neh637c2.pdf.

U.S. Environmental Protection Agency (EPA). 1994. A plain English guide to the EPA part 503 biosolids rule. EPA832-R-93-003. 20 April 2020. http://www.epa.gov/biosolids/plain-english-guide-epa-part-503-biosolids-rule.

U.S. Environmental Protection Agency (EPA). 1995. A guide to the biosolids risk assessments for the EPA part 503 rule. EPA832-B-93-005. 20 April 2020. https://www.epa.gov/biosolids/biosolids-laws-and-regulations.

U.S Environmental Protection Agency (USEPA). 2020. Sustainable management of food. May 17, 2020. https://www.epa.gov/sustainable-management-food/sustainable-management-food-basics#Food%20Waste.

Vaughan R., J.P. Johns, and J. Burkey. 2017. Better technology, better results. *SWS Storm Water Solutions*. 12 June 2020. https://www.estormwater.com/soil-stabilization/better-technology-better-results.

Weathersby, J. 2017. A brief history of golf. *The Sport Historian*. 12 June 2020. https://www.thesportshistorian.com/a-brief-history-of-golf/.

chapter 9

Utilization of compost and food safety

Monica Ozores-Hampton

Fresh fruits and vegetables have been implicated in over 450 outbreaks of foodborne illness in the United States since 1990 (U.S. Department of Agriculture Agricultural Marketing Service, 2018). In commercial food production, it is recommended to establish a risks-reduction strategy using steps outlined in the standards developed by Good Agricultural Practices (GAP), which has been recognized by the Food and Drug Administration (FDA) as the best management practices to foodborne illness-causing pathogens (U.S. Department of Agriculture Agricultural Marketing Service, 2018). The Centers for Disease Control and Prevention (CDC) reported that there are 48 million people who become sick from foodborne illness in the United States every year. However, most people are showing rapid recovery. Nevertheless, there are almost 128,000 hospitalizations and 3,000 deaths yearly caused by foodborne illness, initiated by pathogens (CDC, 2018; Chaifelz et al., 2011).

The Food Safety Modernization Act (FSMA), signed into law in 2011, represents the largest change in food safety in the last 70 years (FDA, 2017; Anonymous, 2017) and regulates the production and uses of compost in food production. The objective of the FSMA is to prevent foodborne illness on farms that grow fresh produce and facilities that process food. The FSMA mandates that measuring compost pile temperatures is the primary and most important factor and method used for pathogen reduction, with the length of time being determined by the composting method (static pile or mechanical turning of the piles) (Bhullar and Andrews, 2015). Bacteria, fungus, parasites and virus pathogens are not easily removed from fresh produce, and therefore limiting contamination from outside inputs is the best practice to reduce foodborne illness (Chaifelz et al., 2011).

Therefore, preventing the introduction of the pathogen in fruit and vegetable production will reduce the risks and increase the food safety of the consumers.

Composting is an ancient practice with its roots dating back thousands of years. Some of the first records can be traced back to a Roman scientist/farmer, Marcus Cato, who lived in Rome at AD 20.

He documented compost to be an "essential additive that ensured soil fertility and productivity." In the early 19th century, composting became the primary means of fertility until the industrial processes were created to produce synthetic chemicals fertilizers in the early 20th century (Ozores-Hampton, 2017b).

Composting is defined as a natural form of recycling organic material which intends to duplicate nature.

The process involves the intake of material or feedstocks from plant and animal origins (complex organic compounds) with an output of the base components as carbon dioxide, water and minerals in a stabilized soil-like material called compost. The transformation process occurs through the action of microorganisms (bacteria, fungi and actinomyces).

As the microorganisms proliferate, they feed and break down the complex organic molecules and generate high temperatures within the compost piles, as the study conducted by Natural Resource, Agriculture, and Engineering Service (NRAES), 1992, recorded temperatures in excess of 165° in commercial composting operations and a reduction of non-beneficial organisms such as human and plant pathogens.

In recent years, composting practices and technologies have progressed remarkably and have been promoted by both government and private agencies involved in waste management. There is a relentless justification for products and practices that are more environment-friendly yet adhere to the strict food safety requirements for consumers of conventional and organic food production in the United States.

A large portion of organic waste can be utilized and reused for agricultural purposes. Properly treated carbon-based waste can be transformed into mature and stable compost (Ozores-Hampton, 2017a). The use of mature and stable compost can improve the agricultural environment by changing the physical, chemical and biological properties of the soil. Applications of compost have been shown to suppress plant pathogens by promoting beneficial microorganisms, further increasing its economic and agricultural impacts (Ozores-Hampton, 2017a, 2012; Ozores-Hampton et al., 2012).

Any carbon-based materials derived from plant or animals can be composted. Animal-based products such as cattle, sheep, horses, swine, poultry, goat, fish and wildlife can harbor pathogen bacteria (*Salmonella and E. coli*), virus and parasites that can grow and spread in the compost, penetrate into the fruits and vegetables and create a food safety risk for the consumer (FDA, 2018; Bhullar and Andrews, 2015) (Table 9.2). Different feedstocks have diverse levels of food safety risk that the composter needs to be aware of during the composting process (Bhullar and Andrews, 2015) (see Table 9.1).

Table 9.1 Summary of the food risk at the farm level using organic amendments

Parameter	Degree of food risk (higher to lower)		
Soil amendment	Raw manure	Composted manure	Chemicals
Crop	Fresh produce		Agronomic
Time of application	Near harvest		Days to harvest
Application methods	Surface		Injected and incorporated
Frequency	Excessive		Adequate

Source: Bhullar and Andrews, 2015.

Table 9.2 Potential animal manure sources and pathogen (bacteria, viruses and parasites) of human disease or illness

Pathogen	Disease/illness	Potential animal manure
Bacteria		
Campylobacter coli and *C. jejuni*	Campylobacteriosis (diarrhea)	Cattle, sheep, swine, poultry, goat and wildlife
Bacillus anthracis	Anthrax (cold and fever)	Cattle, sheep, swine and wildlife
Brucella abortus	Brucellosis (fever)	Cattle, sheep, swine and wildlife
Escherichia coli (pathogenic)	Diarrhea	Cattle, sheep, swine and wildlife
Listeria monocytogenes	Listeriosis (miscarriage in pregnant women)	Cattle
Salmonella spp.	Salmonellosis (diarrhea and fever)	Cattle, sheep, swine, poultry and goat
Yersinia influenza	Yersiniosis	Swine
Virus		
Avian—Swine influenza	Fever, cold, sore throat and diarrhea	Poultry and swine
Hepatitis E	Infects liver (weight loss and nausea)	Swine
Parasites		
Cryptosporidium parvum	Cryptosporidiosis (diarrhea)	Cattle, sheep and swine
Giardia spp.	Giardiasis (diarrhea)	Cattle, sheep and swine
Toxoplasma spp.	Toxoplasmosis (swollen lymph nodes)	Warm-blooded animals

Source: FDA, 2018; Bhullar and Andrews, 2015.

Table 9.3 Food Safety Modernization Act's minimum microbial
standard for composting

Fecal Coliform	Salmonella spp.
Less than 1,000 MPN g^{-1} TS (dry weight)	Not detected using a method that can detect 3 MPN per 4 g^{-1} TS

Source: FDA, 2018; Bhullar and Andrews, 2015.
MPN = most probable number.
TS = total solids.

The temperature of the pile is not uniform during the composting
process. Temperatures are sufficient to kill pathogenic organisms at the
center of the pile or windrow; however, lower temperatures that are insufficient to kill non-beneficial microorganisms occur near the surface of the
pile and require the composting material be turned at specific time intervals. According to the U.S. Environmental Protection Agency (USEPA)
Regulation 503, windrow composting should be operated at a temperature
of at least 131°F for the first 15 days (windrow composting) or 3 days with
in-vessel (static pile) composting methods. For the windrow composting
method, the organic material should be turned five times to eliminate
human and plant pathogens, nematodes and weed seeds (USEPA, 1994,
1995). Therefore, the most powerful strategy to reduce the risk of undesirable bacteria and increase the food safety will be to follow regulations
indicating that compost to be at adequate high temperatures to ensure no
human and plant pathogens remain.

FSMA Produce Safety Rule generated microbial standards for
the thermophilic or temperature-based composting process and vali-
dated them scientifically (FDA, 2018) (Table 9.3). Composting facilities
are required to record these temperatures and maintain the record for
every compost windrow in order to legally use them in food production
(Ozores-Hampton, 2017a).

Separation distance or barriers of the composting operation and crop
production, quality water sources, food contact areas and human hous-
ing should be required to reduce human pathogens (Anonymous, 2017).
Also, compost and animal manure should be stored in separate areas to
decrease cross-contamination by runoff. Traffic and equipment should be
directed away from raw materials to reduce cross-contamination. To prevent the complete elimination of pathogens, regulations that address composting environment, feedstocks inputs, sanitation and storage should be
followed (FDA, 2018; Bhullar and Andrews, 2015; Esser, 1979, 1980, 1984):

a. Improper composting procedures which are frequently caused
   by oversized piles or windrows that cannot be turned properly or
   the lack of uniform temperature distribution in the pile. Also, not

Table 9.4 Optimal desired characteristics for the composting process

Characteristics	Reasonable range	Preferred range
C:N ratio	20:1–40:1	25:1–30:1
Moisture content	40%–65%	50%–60%
Oxygen content	>6%	16%–18.5%
pH	5.5%–%9	6.5–18.5
Bulk density	<40 lb/cubic foot	-
Temperature	110°–140°	130°–140°
Particle size	1/8–2 inches diameter	Varies

addressing the principles of composting such as oxygen requirements, carbon:nitrogen ratio aeration, moisture, porosity, temperature, time and pH (Table 9.4). Unsuitable composting conditions
will result in the failure to eliminate pathogens populations in the
organic raw feedstocks to be composted.

b. Organic raw feedstocks suitable for composting by urban populations include: municipal solid waste; yard trimming; food wastes
from restaurants, grocery stores and institutions; wood wastes from
construction and/or demolition; wastewater (from water treatment
plants); and biosolids (sewage sludge). Similarly, agriculture produces other organic wastes that can be composted: animal manures,
wastes from food processing plants, spoiled feeds and harvest
wastes (Ozores-Hampton et al., 1998, 2005, 2006). However, raw feedstocks require to be blended to produce the correct carbon:nitrogen
ratio for the optimal composting process (Table 9.4).

c. Poor sanitation practices in the compost facility may easily reintro-
duce pathogens in the finished product by recontamination of the
piles during and at the end of the composting process.

d. A site that is not sufficiently elevated and susceptible to be flooded
by water from ponds or canals should also be excluded because these
sites can be readily contaminated with plants or human pathogens
that are transported passively in flood water from infested areas.

e. Improper leveling of the incoming feedstocks receiving area can
allow pathogens movement with flowing water from the receiving
area into the composting and storage areas. It is recommended that
a drainage ditch at least 2-feet deep or a solid wall at least 2-feet high
encompasses any portion of the receiving area facing the composting and storage areas.

f. The selected site should also be protected from strong winds which
are known to transport some pathogens species.

g. Close proximity of the composting process and storage areas
may result in the composting material and stored compost to be

contaminated with raw feedstocks. It is desirable that the incoming raw feedstocks receiving area be established a minimum of 100 feet from the composting and storage areas.

h. The equipment moving the new ground raw feedstock should operate at least 40 feet away from the feedstocks that are already undergoing the composting process.

i. Un-sanitized vehicles can disseminate pathogens during the composting process and after compost. Pathogens can be disseminated by soil and debris clinging to tires and adhering to vehicle parts. Vehicles transporting raw feedstocks to the composting facility and to the hammer mill or to the receiving composting area to initiate the composting process may also be contaminated with pathogens which adhere to debris clinging to the vehicles and thus readily spread to the composting and storage areas. It is desirable that precautionary measures be taken to avoid the entry of un-sanitized machinery into areas with ongoing composting and into the storage area. If possible, equipment should be divided into groups according to the stage of the composting process. There should be one set of equipment used to handle and move only raw feedstocks and another set to handle and move the composted finished products. The equipment (bucket loaders) used to handle raw feedstocks and finished compost should never be in close proximity to one another nor should their working area overlap.

j. There should be an intermediate equipment used only for the turning and aeration of the windrows. Machinery used for turning and aerating the windrows (windrow turners) should start with the oldest (composted) windrows and work toward the newest (raw) feedstock windows in order to prevent contamination of the composted product with the fresh and infected with the finished compost. Once the turning equipment has completed the cycle from the oldest (composted) material to newest (raw) material, it should be thoroughly decontaminated by washing before it returns to the oldest (composted) material. The high temperature of 131°F as an average during the first 15 days and turning of the material five times during the composting process are sufficient to kill any plant and human pathogens. Only through recontamination can these pathogens be reintroduced into the windows or piles. Therefore, the turning equipment should be kept properly sanitized.

k. Irrigation water should come from wells or from ponds free of pathogens to provide moisture during the composting process.

The establishment of a well-designed composting facility and the implementation of appropriate composting and sanitation procedures as defined by FSMA and FDA are both costly and labor intensive. However, these practices will be ensuring the production of high-quality, human and plant pathogen and weed-free compost products to the agricultural industry while ensuring food safety to the consumers.

References

Anonymous. 2017. Food safe compost use. Michigan State University Extension. 17 April 2020. https://www.canr.msu.edu/agrifood_safety/uploads/files/027_01%20reformat-amended%20compost%20use.pdf.

Bhullar, M. and S. Andrews. 2015. FSMA compliant on-farm thermophilic composting: A safe way to enrich the soil. *North Central Regional Center for FSMA Training Extension and Technical Assistance.* 17 April 2020. https://ucanr.edu/sites/Small_Farms_/files/306413.pdf.

Centers for Disease Control and Prevention (CDC). 2018. National outbreak reporting system (NORS). 19 April 2020. https://wwwn.cdc.gov/norsdashboard/.

Chaifelz, A., C. Driscoll, C. Gunter, D. Ducharme, and B. Chapman. 2011. A handbook for beginning and veteran garden organizers: how to reduce food safety risks. *North Carolina Cooperative Extension.* 17 April 2020. https://ucfoodsafety.ucdavis.edu/sites/g/files/dgvnsk7366/files/inline-files/157441.pdf.

Environmental Protection Agency. 1994. A plain English guide to the EPA part 503 biosolids rule. EPA832-R-93-003. 3 February 2016. http://www.epa.gov/biosolids/plain-english-guide-epa-part-503-biosolids-rule.

Esser, R.P. 1979. Nematode entry and dispersion by water in Florida nurseries. Fla. Dept. Agric. & Consumer Serv., Div. Plant Ind., Nema. Circ. No. 54. 2 pp.

Esser, R.P. 1980. Nematode entry and dispersion by man and animals in Florida nurseries. Fla. Dept. Agric. & Consumer Serv., Div. Plant Ind Nema. Circ. No. 60. 2 pp.

Esser, R.P. 1984. How nematodes enter and disperse in Florida nurseries via vehicles. Fla. Dept. Agric. & Consumer Serv., Div. Plant Ind Nema. Circ. No. 109. 2 pp.

Food and Drug Administration (FDA). 2017. FDA food safety modernization act. 19 April 2020. http://www.fda.gov/Food/GuidanceRegulation/FSMA/default.htm.

Food and Drug Administration (FDA). 2018. Produce safety standards. 19 April 2020. https://www.fda.gov/food/guidanceregulation/fsma/ucm304045.htm.

Natural Resource, Agriculture, and Engineering Service (NRAES). 1992. *On-Farm Composting Handbook* (R. Lynk, ed.). 186 pp. Ithaca, NY: NRAES, Cooperative Extension, 14853–5701.

Ozores-Hampton, M. 2006. Soil and nutrient management: compost and manure. In *Grower's IPM Guide for Florida Tomato and Pepper Production,* J.L. Gillett, H.N. Petersen, N.C. Leppla, and D.D. Thomas, 36–40. Gainesville, FL: University of Florida.

Ozores-Hampton, M. 2012. Developing a vegetable fertility program using organic amendments and inorganic fertilizers. *HortTechnology* 22:743–750.

Ozores-Hampton, M. 2017a. Guidelines for assessing compost quality for safe and effective utilization in vegetable production. *HortTechnology* 27:150.

Ozores-Hampton, M. 2017b. Past, present, and future of compost utilization in horticulture. *Acta Horticulturae.* International symposium on growing media, soilless cultivation, and compost utilization in horticulture. 27 Nov. 2020 *https://www.actahort.org/books/1266/1266_43.htm.*

Ozores-Hampton, M., P. Roberts, and P.A. Stansly. 2012. Organic pepper production. In *Peppers: Botany, Production and Uses*, edited by V. Russo, 165–174. Cambridge, MA: CABI.

Ozores-Hampton, M.P., T.A. Obreza, and G. Hochmuth. 1998. Composted municipal solid waste use on Florida vegetable crops. *HortTechnology* 8:10–17.

Ozores-Hampton, M.P., P.A. Stansly, R. McSorley, and T.A. Obreza. 2005. Effects of long-term organic amendments and soil solarization on pepper and watermelon growth, yield, and soil fertility. *HortScience* 40:80–84.

U.S. Environmental Protection Agency (USEPA). 1994. A plain English guide to the EPA part 503 biosolids rule. EPA832-R-93-003. 20 April 2020. http://www.epa.gov/biosolids/plain-english-guide-epa-part-503-biosolids-rule.

U.S. Environmental Protection Agency (USEPA). 1995. A guide to the biosolids risk assessments for the EPA part 503 rule. EPA832-B-93-005. 20 April 2020. https://www.epa.gov/biosolids/biosolids-laws-and-regulations.

United States Department of Agriculture Agricultural Marketing Service. 2018. USDA aligns harmonized GAP program with FDA food safety rule (USDAARS). 19 April 2020. https://www.ams.usda.gov/sites/default/files/media/FAQsUSDAGAPFSMAProduceSafetyRuleAlignment.pdf.

chapter 10

Using compost on nematodes management in horticulture crops

Monica Ozores-Hampton

Contents

Nematodes are responsible for approx. 14% of all worldwide plant losses or $10 billion in the United States and $125 billion annually (Mitkowski and Abawi, 2011). Nematodes belong to the phylum Nematoda and can be defined as ubiquitous pseudocoelomates with unsegmented bodies found in fresh and salt water and soil and as internal parasites of living organisms such as animals and humans (Blaxter, 2011). There are 16–20 different orders within the phylum Nematoda. However, only ten of these orders belong to the soil, and the most common are: Rhabditida, Tylenchida, Aphelenchida and Dorylaimida. Soil nematodes can be classified according to their feeding habits in the soil webs such as herbivores (plant-parasitic nematodes) and free-living nematodes including bacterivores, fungivores, predators and omnivores (Barker et al., 1976). The herbivores nematodes have a mouthpart with a stylet which punctures cells during feeding, either ectoparasites feeding in root surfaces or entoparasites entering the roots to live and feed inside of the roots. Until recently, over 25,000 nematode species have been identified (Zhang, 2013), from which 10,000 have been described as plant-parasites (Maggenti, 1981). Free-living nematodes are most abundant nematodes which feed in microorganisms in the environment in contrast to plant-parasitic nematodes that need a host to feed and survive. Free-living nematodes feed on fungi, algae, fecal matter, plant debris, dead organisms and decomposed soil organic matter (SOM). They play a fundamental role in the decomposition of the SOM and nutrient recycling in the ecosystem (Yeates et al., 1993).

Plant-parasitic nematodes have a life cycle that includes egg stage, four juvenile stages and one adult stage. According to Tyler (1933), for example, a root-knot nematodes (RKN) mature female that establishes a specialized feeding site within plant roots will lay between 500 and 1,000 eggs during her lifetime. The eggs undergo embryogenesis with a first-stage juvenile developed within each egg. The juvenile undergoes the first molt within the eggshell and breaks free of the shell moving into the surrounding soil. After hatching, RKNs are categorized as infectious second-stage juveniles (J2). Root-knot nematodes will infect plant root tips, penetrating in cell elongation of root growing points and will migrate within the roots, wandering around until the movement becomes more purposeful. Root-knot nematodes, with their head oriented toward the region of the plant vascular system, begin to initiate feeding by secreting proteins from the esophageal glands, transforming parasitized cells into hypertrophic cells, also known as "giant cells" (Bird, 1961). The surrounding cortex cell will become hyperplastic, thereby rapidly increasing in numbers to induce gall formation. Hypertrophic cells act as nutrient reservoirs for females. With large numbers of females feeding from inside the roots, this results in nutrient losses and water uptake malfunction, thereby stressing the plant and causing aboveground symptoms such as chlorosis, stunting and incipient wilting of the leaves (Karssen et al., 2013).

Root-knot nematodes, *Meloidogyne* spp. and sting nematodes (*Belonolaimus longicaudatus*) are among the most common and economically relevant plant-parasitic nematode species affecting crops (Mitkowski and Abawi, 2011). Root-knot nematodes can be considered as a main restraining factor for crop production, causing a serious decrease of yield and fruit quality. Root-knot nematodes have indeed been regarded as the greatest nematode threat on multiple crop production in the world (Mitkowski and Abawi, 2011). Plant injury due to RKNs infection involves the formation of "galls" causing poor root function in water and nutrient uptake. Hence, plants exhibit slow recovery to soil moisture adjustments, showing wilting and stunting. Similarly, plants present symptoms distinctive of nutrient deficiency such as chlorosis, ultimately causing yield reduction or losses. Furthermore, RKN-affected plants occur in patches across the field due to the random distribution of RKN population densities (Duncan and Noling, 1998). Besides causing severe, direct crop damage, RKNs also interact with fungi and bacteria, creating plant disease complexes and contributing to yield declines and losses (Mai and Abawi, 1987).

10.1 Using compost on nematode management

Composting is a biological decomposition process in which microorganisms convert organic materials into a relatively stable humus-like material. During decomposition, microorganisms assimilate complex organic substances and release inorganic nutrients (Metting, 1993). An adequate composting process should kill pathogens and stabilize organic carbon (C) before the material is applied to land (Chaney, 1991). New technology and development of processed solid waste materials have resulted in products of high quality to be used by the horticulture industry (Ozores-Hampton, 2012). From an urban viewpoint, compost production represents a safe disposal method for thousands of tons of waste materials produced every year, but the current official recommendations for compost use in horticulture are very general (Ozores-Hampton et al., 2012). Development of alternative production systems for horticultural crops that are environmentally "friendly" yet maintaining optimum yields are needed (Ozores-Hampton et al., 2011). Composts products made from waste materials may provide a significant role in these alternative systems. Compost application may improve soil quality and enhance the utilization of fertilizer, thus improving the performance of fruit, vegetables and ornamental crops (Ozores-Hampton et al., 1998, 2011; Ozores-Hampton and Peach, 2002). Also, compost application may control weeds (Ozores-Hampton et al., 2001a,b), suppress crop diseases (Hoitink and Fachy, 1986; Hoitink et al., 2001), increase SOM, decrease erosion by water and wind (Tyler, 2001) and reduce nutrient leaching (Jaber et al., 2005; Yang et al., 2007).

Numerous studies indicate that applications of compost enhanced crop growth and yields; however, their effect on plant-parasitic nematodes may have a direct nematicidal effect or produce the opposite effects by increasing their population levels. Therefore, compost applications' positive effects on yield may not be directly understood by connecting it to a decrease in plant-parasitic nematodes (Table 10.1). There were 26 studies made assessing the effect of compost on plant-parasitic nematodes on three crop categories that included vegetables (18), fruits (7) and ornamentals (1). Most of the studies concentrated on compost effects on root-lesion *Pratylenchus* spp., root-knot *Meloidogyne* spp., cyst *Globodera* spp. and stubby-root *Paratrichodorus* spp. nematodes. The most common compost tested were municipal solid waste, yard waste, food waste, animal manure, biosolids and by-products from the agriculture industry.

10.2 Nematicidal effect of compost on plant-parasitic nematodes

There were 15 studies that reported a decrease in plant-parasitic nematodes when compost was applied to crop production (Table 10.1).

Table 10.1 The effects of compost type and rate on plant-parasitic nematodes on nuts, fruit and vegetable crop growth and yields

Crop type	Compost type	Compost rate (ton/acre)	Nematode response	Crop response	Reference
			Vegetable crops		
Potatoes	Cull potatoes; sawdust; beef manure	6.5	No effect on *Meloidogyne* (M) *hapla*. Increased on *Pratylenchus* (P) *penetrans*	Increased yield by 27%	Kimpinski et al., 2003
Okra	Neem; cassava peel; sawdust; tithonia	0 and 0.4	Reduced*Heterodera* and *Tylenchus*	Increased plant growth and yield and reduced root damage	Olabiyi and Oladeji, 2014
Peppers/ tomatoes	Olive pomace; PM; MSW	NA	Reduced *M. javanica*	Increased yields	Joan et al., 1997
Sweet corn	YW	0 and 109 as mulch and incorporated	No effect on *M. incognita* and reduced *Criconemella* and *P. spp.*	Increased yields	McSorley and Gallaher, 1996
Yellow squash/okra	YW	0 and 109 as mulch and incorporated	Reduced *M. incognita* and *Criconemella* and no effect on *Paratrichodorus minor, P. spp.* and *Xiphinema spp.*	Increased yields	McSorley and Gallaher, 1995
Tomato	Cattle manure; grape marc	0%, 10%, 25% and 50% (v/v) pots	Reduced *M. javanica* root galls	Increased plant growth	Oka and Yermiyahu, 2002
Potato	Horse; swine manure	0%, 1%, 2.5% and 5% (v/v) pots	Reduced *Globodera* (G) *rostochiensis* and *pallida*	NA	Renco et al., 2011
Potato	YW; PM; BS	0%–10% (w/w)	Reduced *G. rostochiensis*	NA	Renco et al., 2007
Tomato	YW	35%–45% (v/v)	Reduced *M. hapla*	No effect on plant growth	Lozano et al., 2009
Potato/ cucumber	PM	1.2–14.1	Reduced *P. penetrans*	Increased yields	Everts et al., 2006

(Continued)

Table 10.1 (*Continued*) The effects of compost type and rate on plant-parasitic nematodes on nuts, fruit and vegetable crop growth and yields

Crop type	Compost type	Compost rate (ton/acre)	Nematode response	Crop response	Reference
Potato	SM	0 and 6.1	Increased *P. penetrans*	Increased yields by 10%	LaMondia, 2006
No crop	YW	8.1	No effect *P. spp.*	NA	Korthals et al., 2005
Potato/onions	PM	NA	Reduced *P. terest*	NA	Hartsema et al., 2005
Cabbage	Garden waste; farm waste	1.7 inches incorporated	Increased herbivores nematodes	NA	Leroy et al., 2009
Tomato	DM	9 lb/acre N as compost	Decreased plant-parasitic nematodes	NA	Nahar et al., 2006
Tomato	Cotton gin trash	33	No effect on *M. incognita, P.* and reduced *Trichodoridea* and *M. javanica*	NA	Bulluck III et al., 2002
Tomato	Dry cork; grape marc	0%–100% (v/v)	Reduced *M. incognita* and *javanica* population and galling	NA	Nico et al., 2004
Sweet corn beans	PM	4.5	Reduced *Paratrichodorus minor* and no effects on *M. incognita* and *Helicotylenchus dihystera*	No effect	Sumner et al., 2002
Nuts and fruit crops					
Apple	YW; BS	16.1 as mulched	Increased *P. penetrans* soil and roots	Increased root growth	Forge et al., 2008
Apple	BS	4.9–18.2 as mulched	No effects in *P. penetrans*	Increased root growth	Forge et al., 2003

(*Continued*)

Table 10.1 (Continued) The effects of compost type and rate on plant-parasitic nematodes on nuts, fruit and vegetable crop growth and yields

Crop type	Compost type	Compost rate (ton/acre)	Nematode response	Crop response	Reference
Apple	YW mix DM	55.8	No effect on *M.* and *P.*	No effect in tree growth	Yao et al., 2006
Olive	Dry cork; grape marc	0%–100% (v/v)	Reduced *M. incognita* and *javanica*	NA	Nico et al., 2004
Orange	MSW	8	NA	Increased yield and fruit size	Tarjan, 1977
Lemon	MSW	8 (pots)	NA	Increased plant growth and vigor	Tarjan, 1977
Apple	DM	18.2 as mulched	No effect on *P. penetrans*	Slight effect in vigor and yield	Forge et al., 2013
Ornamentals					
Gladiolus	MSW; SM (as mulch)	4.8	No effect on *Trichodoridea*	NA	Zoon et al., 2002

BS = biosolids; MSW = municipal solid waste; DM = dairy manure; PM = poultry manure; SM = spent mushrooms; YW = yard waste; NA = not available.

The nematode-suppressive effects of compost were "directly" attributed to nematotoxic compounds such as high electrical conductivity (EC), high nitrogen (N) concentrations, especially N-ammonia (N-NH$_3$), or accumulation of organic acids such as acetic, propionic and butyric acids (Oka and Yermiyahu, 2002; Thoden et al., 2011). Low C:N ratio composts may accumulate high concentration of N-NH$_3$, being crucial to the development of compost with nematicide effect to control plant-parasitic nematodes. The behavior and effectiveness of compost as a nematicide product based on organic acids and N-NH$_3$ will depend on the soil physico-chemical characteristics such as pH, which has been more effective in acid than neutral or alkaline soils. However, N-NH$_3$ toxicity against plant-parasitic nematodes will perform better in higher pH, since at lower pH it will convert into N-ammonium (N-NH$_4$) (Thoden et al., 2011).

An "indirect" effect of compost application was that it modifies the physico-chemical soil properties that may negatively disturb plant-parasitic nematode hatching, host finding or nematode mobility (Thoden et al., 2011). Also, compost application modifies SOM, improving the physical properties by decreasing bulk density and increasing available water holding capacity (AWHC) and improving soil chemical properties by increasing cation exchange capacity (CEC), pH and macro and micronutrient supplies that enhances plant root and biomass growth, generating resistance to plant-parasitic nematodes (Ozores-Hampton et al., 2011; Sikora and Szmidt, 2001). Another indirect effect of compost application is that it enhances biological properties by increasing soil microbial activity properties of antagonistic and predators bacteria, fungal and nematodes against plant-parasitic nematodes (Thoden et al., 2011; Ozores-Hampton et al., 2011).

10.3 Compost application may increase plant-parasitic nematodes?

There were four studies reported that plant-parasitic nematodes levels increased and six reported that there was no effect on plant-parasitic nematodes after compost application (Table 10.1). According to Thoden et al. (2011), the explanation may be that the high nutrient N-P-K at the root zone may have stimulated fecundity and survivorship, increasing the root-feeding nematodes. Additionally, these environments increased root and plant biomass offering new feeding nematode sites. Crops growing in a high, rich nutrient area reduced the production of secondary metabolites, thereby decreasing self-defenses and increasing the vulnerability of crops to pathogens. Mature and stable compost offers less nematotoxic to the soil since a substantial part of the decomposition process was achieved during the composting process rather than before (immature and unstable compost) (Ozores-Hampton et al., 2011, 2012; Ozores-Hampton, 2017).

The compost mode of action against plant-parasitic nematodes may be the stimulation and production of competitive and antagonistic microorganisms and the proliferation of beneficial plant growth-promoting rhizobacteria and changes on physico-chemical soil properties that promote positive plant growth and development, ultimately increasing the yields (Oka, 2010; Thoden et al., 2011).

10.4   Free-living nematodes and improvement of crop yields the "new paradigm"

Changes in community of free-living nematodes were reported in nine studies indicating that compost application increases non-parasitic, free-living nematodes, especially bacterivores and fungivores (Table 10.2). Additionally, these studies described positive effects on soil microbial populations, SOM, nutrient availability, basal index, channel index, enrichment and the ecosystem stability responsible for the increase of crop yields. Compost application with lower carborn:nitrogen (C:N) ratio will support bacterivores-feeding nematodes and high C:N ratio will benefit fungivores-feeding nematodes. Bacterivores-feeding nematodes returned the nutrients to the soil creating soil-rich nutrient sources instead of being immobilized in their own bodies. Buchan et al. (2013) indicated that in the N mineralization, the activity of nitrifying microorganisms was stimulated by free-nematodes grazing. Additionally, free nematodes promoted the colonization of rhizobacteria, nematode antagonistic bacteria and fungi, together with positive influence on root growth and morphology, modified microbial community and plant hormones.

Compost can be used in conventional and organic fruit, vegetable and ornamental production. Feedstock for composting can be generated from organic-based carbon by-products from urban or agricultural areas. Compost can directly affect soil physical, chemical and biological properties. Therefore, long-term compost improves soil health due to SOM application, reducing the impact on plant stressors such as plant-parasitic nematode, creating an "antifragile" plant farm/soil ecosystem by increasing free-living nematodes diversity, species richness and plant root/plant biomass and crop yields.

Table 10.2 The effects of compost type and rate on free-living nematodes on nuts, fruit, vegetables and turfgrass crop growth and yields

Crop type	Compost type	Compost rate (ton/acre)	Nematode response	Crop response	Reference
			Vegetable crops		
Potato	Cull potatoes; sawdust; beef manure	6.5	Increased free-living	Increased yield by 27%	Kimpinski et al., 2003
Tomato	0, YW; YW/spent mushroom (SM); SM	20%–40% (v/v)	Increased free-living and microbivorous	Reduced yield and increased shelf life	Zhai et al., 2009
Cabbage	Garden waste; farm waste	1.7 inches incorporated	Increased fungivorous *spp.*	NA	Leroy et al., 2009
Tomato	DM	9 lb/acre N as compost	Increased enrichment opportunistic, decreased Shannon-Wiener and MI. Strong negative correlation between total number of free-living and plant-parasitic	NA	Nahar et al., 2006
Tomato	Cotton gin trash	33	Increased *Rhabditidae, Cephalobidae* and fungivorous	NA	Bulluck III et al., 2002
			Nuts and fruit crops		
Apple	YW; BS	16.1 as mulched	Reduced SI and CI and increased bacterial feeders and enrichment opportunists	Increased root growth	Forge et al., 2008

(*Continued*)

Table 10.2 (Continued) The effects of compost type and rate on free-living nematodes on nuts, fruit, vegetables and turfgrass crop growth and yields

Crop type	Compost type	Compost rate (ton/acre)	Nematode response	Crop response	Reference
Apple	DM	18.2 as mulched	Not consistently increased microbivorous	Slight effect in vigor and yield	Forge et al., 2013
Apple	YW mix DM	55.8	Increased free-living	No effect in tree growth	Yao et al., 2006
Ornamentals					
Carnations	Flower-tea	121	Increased free-living		Langat et al., 2008

BS = biosolids; DM = dairy manure; YW = yard waste; MI = maturity index; CI = channel index; SI = structure index; NA = not available.

References

Barker, K.R., P.B. Shoemaker, and L.A. Nelson. 1976. Relationships of initial population densities of Meloidogyne incognita and M. hapla to yield of tomato. *J. Nematol.* 8:232–239.

Bird, A.F. 1961. The ultrastructure and histochemistry of a nematode-induced giant-cell. *J. Biophys. Biochem. Cytol.* 11:701–715.

Blaxter, M. 2011. Nematodes: the worm and its relatives. *Plos. Biol.* 4:9.

Buchan, D., M.T. Gebremikael, N. Ameloot, S. Sleutel, and S. De Neve. 2013. The effect of free-living nematodes on nitrogen mineralisation in undisturbed and disturbed soil cores. *Soil Biol. Biochem.* 60:142–145.

Bulluck III, L.R., K.R. Barker, and J.R. Ristaino. 2002. Influences of organic and synthetic soil fertility amendments on nematode trophic groups and community dynamics under tomatoes. *Appl. Soil Ecol.* 21:233–250.

Chaney, R.L. 1991. Land application of composted municipal solid waste: Public health, safety and environmental issues. *Proc. Natl. Conf. Solid Waste Composting Council*, Falls Church, VA, 13–15 Nov. 1991.

Duncan, L.W., and J.W. Noling. 1998. Agricultural sustainability and nematode IPM. In *Plant-Nematode Interactions*. Chapter 13, edited by K.R. Barker, G.A. Pederson, and G.L. Windham, 251–287. Monograph Series. Madison, WI: Agronomy Society of America.

Everts, K.L., S. Sardanelli, R.J. Kratochvil, D.K. Armentrout, and L.E. Gallagher. 2006. Root-knot and root-lesion nematode suppression by cover crops, poultry litter, and poultry litter compost. *Plant Dis.* 90:487–492.

Forge, T., G. Neilsen, E. Hogue, and D. Faubion. 2013. Composted dairy manure and alfalfa hay mulch affect soil ecology and early production of "Braeburn" apple on M.9 rootstock. *HortScience* 48:645–651.

Forge, A., E.J. Hogue, G. Neilsen, and D. Neilsen. 2003. Effects of organic mulches on soil microfauna in the root zone of apple: implications for nutrient fluxes and functional diversity of the soil food web. *Appl. Soil Ecol.* 22:39–54.

Forge, A., E.J. Hogue, G. Neilsen, and D. Neilsen. 2008. Organic mulches alter nematode communities, root growth and fluxes of phosphorus in the root zone of apple. *Appl. Soil Ecol.* 39:15–22.

Hartsema, O.H., P. Koot, L.P.G. Molendijk, W. Van Den Berg, M.C. Plentinger, and J. Hoek. 2005. *Rotatie onderzoek Paratrichodorus teres*. Praktijkonderzoek Plant en Omgeving, Wageningen UR, The Netherlands. June 1, 2020. https://edepot.wur.nl/120339.

Hoitink, H.A., and P.C. Fachy. 1986. Basis for the control of soil-borne plant pathogens with composts. *Annu. Rev. Phytopathol.* 24(1):93–114.

Hoitink, H.A., M.S Krause, and D.Y. Han. 2001. Spectrum and mechanisms of plant disease control with composts. In *Compost Utilization in Horticultural Cropping Systems*, edited by P.J. Stoffella, and B.A. Kahn, 263–274. Boca Raton, FL: CRC Press.

Jaber, F.H., S. Shukla, P.J. Stoffella, T.A. Obreza, and E.A. Hanlon. 2005. Impact of organic amendments on groundwater nitrogen concentrations for sandy and calcareous soils. *Compost Sci. Util.* 13(3):194–202. doi: 10.1080/1065657X.2005.10702240.

Joan, M., J. Pinochet, and R. Rodriguez-Kabana. 1997. Agricultural and municipal compost residues for control of root-knot nematodes in tomato and pepper. *Compost Sci. Util.* 5(1):6–15.

Karssen, G., W. Wesemael, and M. Moens. 2013. Root-knot nematodes. In *Plant Nematology*, 2nd ed., edited by R.N. Perry and M. Moens, 73–108. Wallingford, UK: CAB International.

Kimpinski, J., C.E Gallant, R. Henry, J.A Macleod, J.B. Sanderson, and A.V. Sturz. 2003. Effect of compost and manure soil amendments on nematodes and on yields of potato and barley: a 7-year study. *J. Nematol.* 35:289–293.

Korthals, G.W., J.H.M. Visser, and L.P.G. Molendijk. 2005. Improvement and monitoring soil health. *Acta Hortic.* 698:279–284.

Lamondia, J.A. 2006. Management of lesion nematodes and potato early dying with rotation crops. *J. Nematol.* 38:442–448.

Langat, J.K., J.W. Kimenju, G.K. Mutua, W.M. Muiru, and W. Otieno. 2008. Response of free-living nematodes to treatments targeting plant-parasitic nematodes in carnation. *Asian J. Plant Sci.* 7:467–472.

Leroy, B.L.M.M., N. De Sutter, H. Ferris, M. Moens, and D. Reheul. 2009. Short-term nematode population dynamics as influenced by the quality of exogenous organic matter. *Nematology* 11:23–38.

Lozano, J., W.J. Block, and A.J. Termorshuizen. 2009. Effect of compost particle size on suppression of plant diseases. *Environ. Eng. Sci.* 26:601–607.

Maggenti, A. 1981. *General Nematology.* New York: Springer-Verlag.

Mai, W.F., and G.S. Abawi. 1987. Interactions among root-knot nematode and fusarium wilt fungi on host plants. *Ann. Rev. Phytopathol.* 25:317–38.

McSorley, R., and R.N. Gallaher. 1995. Effect of yard waste compost on plant-parasitic nematode densities in vegetable crops. *J. Nematol.* 27(4S):545–549.

McSorley, R., and R.N. Gallaher. 1996. Effect of yard waste compost on nematode densities and maize yield. *J. Nematol.* 28(4S):655–660.

Metting, F.B. 1993. *Soil Microbial Ecology. Application in Agricultural and Environmental Management.* New York: Marcel Dekker, Inc.

Mitkowski, N.A., and G.S. Abawi. 2011. Root-knot nematode. *The Plant Health Instructor.* June 22, 2020. https://www.apsnet.org/edcenter/disandpath/nematode/pdlessons/Pages/RootknotNematode.aspx.

Nahar, M.S., P.S. Grewal, S.A. Miller, D. Stinner, B.R. Stinner, M.D. Leinhenz, A. Wszelaki, and D. Doohan. 2006. Differential effects of raw and composted manure on nematode community, and its indicative value for soil microbial, physical and chemical properties. *Appl. Soil Ecol.* 34:140–151.

Nico, A., R. Jimenez-Diaz, and P. Castillo. 2004. Control of root-knot nematodes by composted agro-industrial wastes in potting mixtures. *Crop Protect.* 23(7):581–587.

Ozores-Hampton, M. 2012. Developing a vegetable fertility program using organic amendments and inorganic fertilizers. *HortTechnology* 22:743–750.

Ozores-Hampton, M. 2017. Impact of soil health and organic nutrient management on vegetable yield and quality. *HortTechnology* 27:162–165.

Ozores-Hampton, M., T.A. Obreza, and P.J. Stoffella. 2001a. Weed control in vegetable crops with composted organic mulches. In *Compost Utilization in Horticultural Cropping Systems*, edited by P.J. Stoffella and B.A. Kahn, 275–286. Boca Raton, FL: CRC Press.

Ozores-Hampton, M., P. Roberts, and P.A. Stansly. 2012. Organic pepper production. In *Peppers: Botany, Production and Uses*, edited by V. Russo, 165–174. Cambridge, MA: CABI.

Ozores-Hampton, M., P.A. Stansly, and T. Salame. 2011. Soil chemical, physical, and biological properties of a sandy soil subjected to long-term organic amendments. *J. Sustain. Agr.* 35:243–259.

Ozores-Hampton, M.P., T.A. Obreza, and G. Hochmuth. 1998. Composted municipal solid waste use on Florida vegetable crops. *HortTechnology* 8:10–17.

Ozores-Hampton, M.P., T.A. Obreza, P.J. Stoffella, and G. Fitzpatrick. 2001b. Immature compost suppresses weed growth under greenhouse conditions. *Compost Sci. Util.* 10(2):105–113. doi: 10.1080/1065657X.2002.10702071.

Ozores-Hampton, M.P., and D.R. Peach. 2002. Biosolids in vegetable production systems. *HortTechnology* 12:8–22.

Oka, Y. 2010. Mechanisms of nematode suppression by organic soil amendments – A review. *Appl. Soil Ecol.* 44:101–115.

Oka, Y., and U. Yermiyahu. 2002. Suppressive effects of composts against the root-knot nematode Meloidogyne javanica on tomato. *Namatology* 4(8):891–898.

Olabiyi, T.I., and O.O. Oladeji. 2014. Assessment of four compost types on the nematode population dynamics in the soil sown with okra. *Int. J. Org. Agric. Res. Dev.* 9(146):146–155.

Renco, M., T. D'addabbo, N. Sasanelli, and I. Papajova. 2007. The effect of five composts of different origin on the survival and reproduction of *Globodera rostochiensis*. *Nematology* 9:537–543.

Renco, M., H. Sasanellin, and P. Kovacok. 2011. The effect of soil compost treatments on potato cyst nematodes Globodera rostochiensis and Globodera pallida. *Helminthologia* 48(3):184–194.

Sikora, L.J., and R.A. Szmidt. 2001. Nitrogen sources, mineralization rates, and nitrogen nutrition benefits to plants from composts. In *Compost Utilization in Horticultural Cropping Systems*, edited by P.J. Stoffella and B.A. Kahn, 287–306. Boca Raton, FL: CRC Press.

Sumner, D.R., M.R. Hall, J.D. Gay, G. Macdonald, S.I. Savage, and R.K. Bramwell. 2002. Root diseases, weeds, and nematodes with poultry litter and conservation tillage in a sweet corn-snap bean double crop. *Crop Prot.* 21:963–972.

Tarjan, A.C. 1977. Applications of municipal solid waste compost to nematode-infected citrus. *Nematropica* 7(2):53–56.

Thoden, T.C., G.W. Korthals, and A.J. Termorshuizen. 2011. Organic amendments and their influences on plant-parasitic and free-living nematodes: a promising methods for nematode management? *Nematology* 13(2):133–152.

Tyler, J. 1933. Reproduction without males in aseptic root cultures of the root-knot nematode. *Hilgardia* 7:373–388.

Tyler, R. 2001. Compost filter berms and blankets take on the silt fence. *Biocycle* 42(1):26–31.

Yang, J., Z. He, Y. Yang, P.J. Stoffella, X.E. Yang, D.J. Banks, and S. Mishra. 2007. Use of amendments to reduce leaching of phosphate and other nutrients from a sandy soil in Florida. *Environ. Sci. Pollution Res.* 14(4):266–269. doi: 10.1065/espr2007.01.378.

Yao, S., I.A. Merwin, G.S. Abawi, and J.E. Thies. 2006. Soil fumigation and compost amendment alter soil microbial community composition but do not improve tree growth or yield in an apple replant site. *Soil Biol. Biochem.* 38:587–599.

Yeates, G.W., T.Bongers, R.G. Goede, D.W. de Freckman, and S.S. Georgieva. 1993. Feeding habits in soil nematode families and genera - an outline for soil ecologists. *J. Nematol.* 25:315–331.

Zhai, Z., D.L. Ehret, T. Forge, T. Helmer, W. Linorais, M. Dorais, and A.P. Apadopoulus. 2009. Organic fertilizers for greenhouse tomatoes: productivity and substrate microbiology. *HortScience* 44:800–809.

Zhang, Z.Q. 2013. Animal biodiversity: an update of classification and diversity in 2013. *Zootaxa* 3703:5–11.

Zoon, F.C., A.S. Van Bruggen, A. De Heij, C.J. Asjes, and J.E. Van Den Ende. 2002. Effect of green manure crops and organic amendments on incidence of nematode-borne tobacco rattle virus. *Acta Hort.* 598:287–29.

chapter 11

Compost tea foliar disease suppression in horticulture crops

Craig S. Coker and Monica Ozores-Hampton

Contents

Compost teas are watery, fermented, liquid compost extracts (Scheuerell and Mahaffee, 2013). Other terminologies used are compost leaching, compost extract, aerobic compost tea (ACT) or nonaerobic (NCT) or animal manure teas. This concentrate liquid is a source of beneficial microorganism (bacteria, fungi, protozoa and nematodes) macro and micronutrients and microorganisms' metabolites such as humid acids and hormones plant regulators.

Compost teas can be used primarily to increase plant production in horticulture crops as part of the fertility program, improve and remediate soils, increase crop vigor and yields, enhance the composting process, control odors and control soil and foliar diseases (Scheuerell and Mahaffee, 2002). However, compost tea is not a fertilizer, a fungicide, a pesticide or a product that does everything.

11.1 Benefits of compost teas

There are several documented benefits from applying compost teas to soil and/or plants. Specifically, compost teas promote soil biological activity, improve soil structure, increase plant health and vigor, reduce fungicide application and fertilizer application rates and promote diseases suppression together with a minimum impact in the environment (Siddiqui et al., 2015). In the soil, compost teas can increase water retention and improve soil fertility and reduce reliance on the need for chemical pesticides and fertilizers (Joe et al., 2017). Conventional compost also produces benefits to both plants and soils, but, unlike compost tea, has not been acknowledged to reduce plant soil and foliar pathogens. Compost tea applied directly to plant foliage has been showed to suppress phytopathogens on a variety of edible crops, including nuts and fruits, vegetables and ornamental plants (Scheuerell and Mahaffee, 2013). Other benefits include potential for decreased water use (irrigation) due to increased water holding capacity; induced microbial proliferation due to increased labile nutrients; and enhanced microbial activity in the rhizosphere due to optimal water supplies (Joe et al., 2017). Compost teas promote soil biological activity as a source of inoculant to enhance nutrient recycling by beneficial bacteria, fungi, protozoa and nematodes promoting carbon, nitrogen, sulfur and other natural soil nutrient cycles in the soil.

The primary benefit of compost teas is in plant disease suppression, which is believed to be due to the presence of microorganisms in the tea (Evans et al., 2013). Biological interactions that result in disease suppression of plant and soilborne pathogens are complex because diseases caused by pathogens occur in a dynamic environment. These interactions are thought to occur through the following mechanisms, which are not necessarily mutually exclusive (Dearborn, 2011):

- *Antibiosis*: some beneficial organisms can produce antibiotics or other substances that are toxic to the pathogenic organisms. For example, bacteria *Pseudomonas fluorescens* strain CHAO produces hydrogen cyanide, 2,4-diacetylphloroglucinol and pyoluteorin, which directly interfere with the growth of various pathogens. Other bacteria including *Bacillus* and *Serrantia* and fungi such as *Trichoderma* and *Gliocladium* can produce antimicrobial compounds effective against plant root pathogens.
- *Competition*: when beneficial microorganisms are present in a growing medium, they tend to outcompete pathogenic bacteria or fungi for food source.
- Induced resistance: some beneficial microbes colonizing on plant roots or foliage are documented to confer resistance to plant by turning on genes that increase plant tolerance to infection by pathogens.
- *Parasitism*: certain beneficial microbes can feed on specific pathogens. For example, *Trichoderma* spp. are shown in various studies to secrete enzymes that digest the cell wall of some fungal root pathogens.

11.2   Compost teas types

Compost teas can be divided in NCT and ACT (Scheuerell and Mahaffee, 2013). Nonaerobic compost tea is a passive method that is not included active aeration or agitation of the components: compost, water, time and optional nutrients. Similarly, ACT is an active method that includes agitation during the brewing process, however, has the same components as NCT. There is limited scientific data that directly compares compost tea production processes (Evans et al., 2013). However, available data suggests that both ACT and NCT can be inconsistent from batch to batch. The inconsistency has been associated to several factors that affect the production process included in Figure 11.1 (Dearborn, 2011):

Compost grade: composts can be made from animal manures, landscape debris, food wastes, agricultural residuals and biosolids. Some research suggests composition of microorganisms in compost depends on the feedstock. For example, carbon-rich feedstocks (e.g., dry leaves, sawdust, wood chips and shredded newspaper) produce compost with a higher fungal content while nitrogen-rich feedstocks (hay weeds, coffee grounds, herbaceous material and manures) produce compost with higher bacterial content (Scheuerell and Mahaffee, 2006).

Compost-water ratio: the ratio of compost to water (volumetric) varies for each production method (Dearborn, 2011). For NCT, most studies use a 1:3–1:10 ratio (Scheuerell and Mahaffee, 2002). For ACT, the ratio depends on the type of equipment and is usually suggested by the compost tea equipment suppliers.

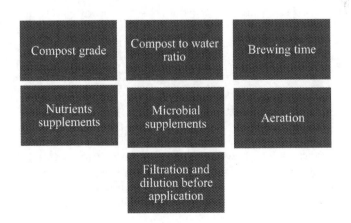

Figure 11.1 Factors that affect the compost tea production process.

Brewing time: several studies on disease suppression properties of NCT have indicated that NCT brewing time of 8–16 days is optimal fermentation time for any level of disease control (Scheuerell and Mahaffee, 2002). An advantage reported by the manufacturers and the users of ACT is that the short brewing time of between 18 hours and 3 days makes the tea readily available (Dearborn, 2011).

Nutrient supplements: nutrients can be added to compost teas to act as catalysts or microbial starters. These include kelp, fish hydrolysate, molasses and humic acid or proprietary additives available from compost tea and brewing equipment manufacturers (Dearborn, 2011). One consideration in the use of molasses or simple sugars would be the potential for proliferation of pathogens if the compost grade is not mature enough (Ingram and Millner, 2007). However, molasses may encourage regrowth of human pathogenic bacteria, creating public concern with potential contamination of fresh crops consumption. Studies found that regrowth of *Salmonella enterica* serovar Thompson and *Escherichia coli* O157:H7 was positively correlated to molasses concentration in dairy and poultry manure by brewing for 72 hours (Duffy et al., 2004).

Microbial supplements: commercial suppliers of compost teas often advertise pre-packaged microbial inoculants that can be brewed independently, added to the compost being brewed or added after the tea is made. These inoculants can be a blend of facultative bacteria,[1] yeast, enzymes, trace minerals, vitamins and organic acids (Dearborn, 2011).

Aeration: as composting is widely defined as the aerobic decomposition of organic matter, some compost tea producers support aerating the tea to promote the growth and propagation of a diverse group of beneficial microbes. In one study, NCT and ACT brewing techniques were compared with and without aeration and in the presence or absence of nutrient additive for suppression of the fungus *Pythium* damping-off of cucumber (*Cucumis sativus.* L.) seedlings. The study showed that no significant correlation could be drawn between the microbial population in the compost tea brewed under continuous aeration and disease suppression (Dearborn, 2011). However, addition of nutrients to ACT during the brewing process showed the most consistent suppression of *Pythium* damping-off, suggesting that nutrient, and not necessarily aeration, support the microbial activity in ACT (Scheuerell and Mahaffee, 2006; Dearborn, 2011).

11.3   Application methods and compost teas quality

Common compost teas application methods included central pivot, micro-jet or drip irrigation, tractors-driven and manual sprayers. Prior to application, it is recommended to establish a minimum compost teas quality control program that included compost and chemical/biological analysis of the compost teas. There are no U.S. government restrictions on how and when compost teas can be used in horticulture production; however, to eliminate or reduce human and plant pathogen, nematodes and weeds, the temperature of the material during the composting process (thermophilic stage) to be used for compost teas must be an average of 131°F for 3 days in an in-vessel or static aerated pile or first 15 days in windrow composting methods and be turned at least five times during this period, according to the Clean Water Act 40 CFR Part 503 (USEPA, 1994, 1995, 1999). Additionally, minimum biological compost and compost teas standards had been established to optimize the finish product to be used in horticulture crops (Table 11.1). Compost teas should be tested for fecal coliform and salmonella to facilitate growers to reduce the risks of contamination to crops.

Compost tea applied can be used as a soil drench (soil pathogens) or as a foliar spray (foliar pathogens) (Diver, 2002). When applied as a soil drench, the recommended application rate is 15 gallons/acre to the soil around the plant (Ingham, 2005). Compost teas can be applied at full strength or diluted at compost with tea:water ratio of 1:10 at 20–40 gallons/acre.

Table 11.1 Minimum compost biological standards to be utilized in compost teas for horticultural crop application

Parameter	Range (g/compost)	Range (ml/compost tea)
Moisture (%)	50–70	–
Active bacteria (ug)	2–10	10–150
Total bacterial (ug)	150–300	150–300
Active fungi (ug)	2–10	2–10
Total fungi (ug)	150–300	5–20
Flagellates	10,000	1,000
Amoebas	19,000	1,000
Ciliates	50–100	20–50
Beneficial nematodes	10–50	2–10

Table source: Diver, S., 2002.

Foliar application rates vary with the size of the plant being treated, with the goal of 70% leaf coverage. For seedlings and small plants, recommended application rates will be 5 gallons/acre. For larger plants with their larger foliage areas, application rates would be 2.6 gallons/acre per 3 feet of plant height (approx. 1 gallon/acre per foot of plant height). Underapplication is considered more problematic than overapplication (Ingham, 2005).

11.4   Use of compost teas to promote foliar diseases suppression in horticulture crops

Due to the increased development of fungicide-resistant isolates, potential regulatory loss of fungicides and worker, health and safety concerns, alternative foliar disease control measures are needed in the organic and conventional horticulture industry (Haggag and Saber, 2007). Compost teas are a biological control method that has potential to suppress a broad range of plant pathogens (Scheuerell and Mahaffee, 2013). However, foliar plant diseases using compost teas are a dynamic complex system; therefore, making recommendations for its use is more complicated than standard fungicides. The lack of knowledge by agricultural professionals' results in failure in the use of compost teas to control plants diseases or mistakes and problems with compost teas production and uses. Foliar application of compost teas will provide the foliage with microorganisms and nutrients in the surface of the leaves essential for diseases suppression.

Compost teas can be used in conventional or organic crop production. In conventional crop production, compost teas can be used when limited fungicide are available or as part of the fungicide rotations program. In organic crop production, chemical fungicides are prohibited, and therefore compost teas can be an essential part of the foliar diseases control program. A large number of foliar diseases can be controlled by compost teas in ornamental, vegetable and nut and fruit crops (Table 11.2). Different combinations of compost tea methods (ACT or NCT), compost feedstocks types, dilution rates, breeding time, nutrient or microbial supplements etc. can control or reduce the severity of the same foliar diseases. Therefore, growers are encouraged to combine the different factors that affect the effectiveness of the compost teas to obtain the optimal foliar diseases control.

Further research is needed to validate strategies of compost teas as a suitable tool to control plant diseases and an end-user guideline needs to be developed to provide a wider use of compost teas in horticulture crops.

Table 11.2 The effects of compost teas on foliar nuts, fruit, vegetables and turfgrass diseases

Crop type	Diseases	Compost type/dilution rate (compost:water)	Disease response	Reference
Ornamentals				
Geranium	Gray mold (*Botrytis cinerea*)	ACT, kelp, rock dust, humid acid and adjuvants	Reduced the disease	Scheuerell and Mahaffee, 2006
Roses	Powdery mildew (*Sphaerotheca pannosa* var. rosae)	ACT 1:8	Suppressed the disease	Seddigh and Kiani, 2018
Roses	Powdery mildew	ACT and NCT; PM, YW and mixed sources	Suppressed the disease	Scheuerell and Mahaffee, 2000
Roses	Powdery mildew	NA	Reduced the disease	Seddigh et al., 2014
Vegetables				
Bean	Gray mold (*Botrytis cinerea*)	ACT	Suppressed disease	Palmer et al., 2010
Bean	Angular leaf spot (*Phaeoisariopsis griseola* (Sacc.) Ferraris)	NCT and PM	Reduced the disease	Joshi et al., 2009
Melon	Gummy stem blight (*Didymella bryoniae*) and powdery mildew (*Podosphaera fusca*)	ACT and NCT, spent mushroom, grape marc, YW and vermicompost	Reduced the disease	Marin et al., 2013
Okra	Wet rot (*Choanephora cucurbitarum*)	NCT, rice straw and oil palm	Reduced the disease	Siddiqui et al., 2009
Onion	Purple blight (*Alternaria porri*)	PM	NCT reduced the disease	Haggag and Saber, 2007
Potato	Scab (*Streptomyces scabiei*)	ACT	Reduced severity	Al-Mughrabi et al., 2008
Pepper and cucumber	Anthracnoses (*Colletotrichum coccodes* and *C. orbiculare*)	NCT	Reduced severity	Sang and Kim, 2011

(Continued)

Table 11.2 (Continued) The effects of compost teas on foliar nuts, fruit, vegetables and turfgrass diseases

Crop type	Diseases	Compost type/dilution rate (compost:water)	Disease response	Reference
Strawberries	Gray mold (Botrytis cinerea)	ACT and NCT 1:8 and 1:4	Reduced severity	Welke, 2005
Tomato	Early blight (Alternaria solani), gray mold (Botrytis cinerea) and powdery mildew (Oidium neolycopersici)	NCT and sheep manure	Suppressed early blight gray mold, but failed powdery mildew	Kone et al., 2010
Tomato	Powdery mildew (Erysiphe polygoni)	ACT and garden waste	Suppressed the disease	Segarra et al., 2009
Tomato	Late blight (Alternaria Septoria)	ACT and vermi-compost	Failed to suppress	Barker-Plotkin, 2000
Tomato	Early blight (Alternaria solani)	NCT and PM	Reduced the disease	Haggag and Saber, 2007
Tomato	Early blight and gray mold (Botrytis cinerea)	NA	Reduced the disease	On et al., 2015
Tomato	Gray mold, black spot and Alternaria alternata	ACT, YW and FC 1:15	Reduced the disease	Pane et al., 2012
Nuts and fruits				
Blueberry	Mummy berry disease (Monilinia vaccinii-corymbosi)	NA	Failed to suppress	McGovern et al., 2012
Cacao	Witches' broom disease (Moniliophthora perniciosa)	NA	Suppressed disease at 2.75%	Maridueña-Zavala et al., 2019
Grapevine	Botrytis bunch rot (Botrytis cinereal) and powdery mildew (Erysiphe necator)	ACT	Reduced severity	Evans et al., 2013
Pecan	Scab (Venturia effuse)	NA	Reduced the disease	Bock et al., 2019

FC = food compost; PM = poultry manure; YW = yard waste; ACT = aerobic compost tea; NCT = nonaerobic compost tea; NA = not available.

Note

1. Bacteria that can use dissolved oxygen (DO) or oxygen obtained from food materials such as sulfate or nitrate ions, or some can respire through glycolysis. The bacteria can live under aerobic, anoxic or anaerobic conditions

References

Al-Mughrabi, K.I., C. Berthélémé, T. Livingston, A. Burgoyne, R. Poirier, and A. Vikram. 2008. Aerobic compost tea, compost and a combination of both reduce the severity of common scab (*Streptomyces scabiei*) on potato tubers. *Journal Plant Science* 3:168–175.

Barker-Plotkin, J. 2000. Biological control of tomato foliar disease. Northeast SARE Project. Number FNE00-292. Burlington, VT.

Bock, C.H., M.W. Hotchkiss, D.I. Shapiro-Ilan, J.H. Brock, T.B. Brenneman, B. Wilkins, D.E. Wells, L. Wells, and R.F. Mizell. 2019. A comparison of organic fungicides: alternatives for reducing scab on pecan. *Organic Agriculture* 9(3):305–314.

Dearborn, Y. 2011. *Compost Tea-Literature Review on Production, Application and Plant Disease Management.* San Francisco Department of Environment Toxic Reduction Program: IPM Task Order 3–18.

Diver, S. 2002. *Notes on Compost Teas.* A Supplement to the Attra publication "compost teas for plant diseases control." Attra March.

Duffy B., C. Sarreal, S. Ravya, and L. Stanker. 2004. Effect of molasses on regrowth of *E. coli* O157:H7 and *Salmonella* in compost teas. *Compost Science & Utilization* 12(1):93–96.

Evans, K.J., A.K. Palmer, and D.A. Metcalf. 2013. Effect of aerated compost tea on grapevine powdery mildew, botrytis bunch rot and microbial abundance on leaves. *European Journal of Plant Pathology* 135(4):661–673.

Haggag, W.M., and M.S.M. Saber. 2007. Suppression of early blight on tomato and purple blight on onion by foliar sprays of aerated and non-aerated compost teas. *Journal of Food, Agriculture & Environment* 5(2):302–309.

Ingham, E. 2005. *The Compost Tea Brewing Manual*, 5th ed. Corvallis, OR: Soil Foodweb, Inc.

Ingram, D.T., and P.D. Millner. 2007. Factors affecting compost tea as a potential source of *Escherichia coli* and *Salmonella* on fresh produce. *Journal Food Protection* 70(4):828–834.

Joe, V., C. Rock, and J. McLain. 2017. *Compost Tea 101: What Every Organic Gardener Should Know.* University of Arizona Cooperative Extension Rpt. Az1739.

Joshi, D., K.S. Hooda, J.C. Bhatt, B.L. Mina, and H.S. Gupta. 2009. Suppressive effects of composts on soil-borne and foliar diseases of French bean in the field in the western Indian Himalayas. *Crop Protection* 28:608–615.

Kone, B.S., A. Dionne, R.J.T. Tweddell, H. Antoun and T.J. Avis. 2010. Suppressive effect of non-aerated compost teas on foliar fungal pathogens of tomato. *Biological Control* 52(2):167–173.

Maridueña-Zavala, M.G., A. Freire-Peñaherrera, R.F. Espinoza-Lozano, M. Villavicencio-Vasquez, M. Jimenez-Feijoo, and J.M. Cevallos-Cevallos. 2019. Genetic characterization of *Moniliophthora perniciosa* from Ecuador and in vitro sensitivity to compost tea. *European Journal of Plant Pathology* 154(4):943–959.

Marín, F., M. Santos, F. Diánez, F. Carretero, F.J. Gea, J.A. Yau, and M.J. Navarro. 2013. Characters of compost teas from different sources and their suppressive effect on fungal phytopathogens. *World Journal of Microbiology and Biotechnology* 29(8):1371–1382.

McGovern, K.B., S.L. Annis, and D.E. Yarborough. 2012. Efficacy of organically acceptable materials for control of mummy berry disease on lowbush blueberries in Maine. *International Journal of Fruit Science* 12(1–3):188–204.

On, A., F. Wong, Q. Ko, R.J. Tweddell, H. Antoun, and T.J. Avis. 2015. Antifungal effects of compost tea microorganisms on tomato pathogens. *Biological Control* 80:63–69.

Palmer, A.K., K.J. Evans, and D.A. Metcalf. 2010. Characters of aerated compost tea from immature compost that limit colonization of bean leaflets by Botrytis cinerea. *Journal of Applied Microbiology* 109(5):1–13.

Pane, C., G. Celano, D. Villecco, and M. Zaccardelli. 2012. Control of botrytis *cinerea*, *Alternaria alternata* and *Pyrenochaeta lycopersici* on tomato with whey compost-tea applications. *Crop Protection* 38:80–86.

Sang, M.K, and K.D. Kim. 2011. Biocontrol activity and primed systemic resistance by compost water extracts against anthracnoses of pepper and cucumber. *Phytopathology* 101:732–740.

Scheuerell, S.J., and W.F. Mahaffee. 2000. Foliar disease suppression with aerobic and anaerobic watery fermented compost and Trichoderma harzianum T-22 for of powdery mildew (*Sphaerotheca pannosa* var. rosea) of Rose in the Willamette Valley, Oregon (abstract). *Phytopathology* 90s:69.

Scheuerell, S.J., and W.F. Mahaffee. 2002. Compost tea principals and prospects for plant disease control. *Compost Science and Utilization* 10(4):313–338.

Scheuerell, S.J., and W.F. Mahaffee. 2006. Variability associated with suppression of gray mold (*Botrytis cinerea*) on geranium by foliar applications of nonaerated and aerated compost teas. *Plant Disease* 90:1201–1208.

Scheuerell, S., and W. Mahaffee. 2013. Compost tea: Principles and prospects for plant disease control. *Compost Science & Utilization* 10(4):313–338.

Seddigh, S., and L. Kiani. 2018. Evaluation of different types of compost tea to control rose powdery mildew (*Sphaerotheca pannosa* var. rosae). *International Journal of Pest Management* 64(2):178–184.

Seddigh, S., L. Kiani, B. Tafaghodinia, and B. Hashemi. 2014. Using aerated compost tea in comparison with a chemical pesticide for controlling rose powdery mildew. *Archives of Phytopathology and Plant Protection* 47(6):658–664.

Segarra, G., M. Reis, E. Casanova, and I. Trillas. 2009. Control of powdery mildew (*Erysiphe polygoni*) in tomato by foliar applications of compost tea. *Journal of Plant Pathology* 91:683–689.

Siddiqui, Y., S. Meon, R. Ismail, and M. Rahmani. 2009. Bio-potential of compost tea from agro-waste to suppress *Choanephora cucurbitarum* L. the causal pathogen of wet rot of okra. *Biological Control* 49(1):38–44.

Siddiqui Y., Y. Naidu, and A. Asgar. 2015. Bio-intensive management of fungal diseases of fruits and vegetables utilizing compost and compost teas. *Organic Amendments and Soil Suppressiveness in Plant Disease Management* 41:307–329.

U.S. Environmental Protection Agency (USEPA). 1994. *A Plain English Guide to the EPA Part 503 Biosolids Rule*. EPA832-R-93-003. Sept. Washington, DC.

U.S. Environmental Protection Agency (USEPA). 1995. *A Guide to the Biosolids Risk Assessments for the EPA Part 503 Rule.* EPA832-B-93-005. Sept. Washington, DC.

U.S. Environmental Protection Agency (USEPA). 1999. *Biosolids Generation, Use, and Disposal in the United States.* EPA503-R-99-009. Sept. Washington, DC.

Welke, S. 2005. The effect of compost extract on the yield of strawberries and the severity of *Botrytis cinereal. Journal of Sustainable Agriculture* 25(1):57–68.

chapter 12

Compost utilization and the economic impact in crop production

Rodney W. Tyler

Contents

The horticulture industry is well suited to composting and compost. The amount and nature of farm wastes, the availability of land and the benefits which compost brings to soil make farms an ideal place to practice composting (Ozores-Hampton, 2017; Stoffella et al., 2014). Crop production using compost is difficult to measure in many situations because most horticulture operations need to pay for "inputs" into their farm with "outputs" of what they grow and sell. Therefore, most inputs need to be calculated on a value of return basis, which is challenging when using compost because the benefits last for many years. Amortizing the cost of using compost over 5 years makes more sense due to benefits of increased water and nutrient holding capacity, organic matter (OM), drainage and several other cultural benefits. Most farmers are not used to amortizing inputs over long of a period, because horticulture has been addicted to annual inputs of man-made fertilizers. Using compost in combination with other annual inputs needs to be calculated and planned like any other farm activity.

The size of the horticulture market for compost use is nearly incomprehensible. One of horticulture's major drawbacks is the distance from urban compost sites, which can add significant freight costs to the product. However, over the last 30 years, many farms have started composting farm wastes such animal manure, bedding and other organics. Some have converted to creating biogas with their on-farm wastes and then land-applied the residuals. Some have incorporated the use of incoming yard waste (YW) as a bulking agent for animal manures (Rynk et al., 1992). This option creates a potential revenue stream of inputs which farmers are not used to. Since the horticulture market changes yearly in the form of reduced acreage due to urbanization and set-aside government crop programs, this enormous market will continue to gradually shrink into the future. However, from all data available, the horticulture market can easily absorb all compost that we could ever produce from all organic residuals and on-farm organic residuals.

Economic impact for crop production using compost is a very straightforward calculation at first glance. The outputs from farm production should pay for inputs of compost applications. Right? The answer is it depends. The livelihood of productive farm fields is built within the reserves of good quality soils, based on good soil OM (SOM). This chapter focuses on technical aspects of variables associated with these principles, while discussing economic factors that impact decisions on horticulture. Crops such strawberries (*Fragaria vesca* ssp. *americana* (Porter) Staudt) yielding $20–$30,000/acre from production compared to corn (*Zea mays* L.) or soybeans (*Glycine max* L.) at less than $1,000/acre suggest that only high dollar crops may be able to afford compost applications. However, all good farm managers will indicate that without the healthy soil, production of the lowest value commodities will be less profitable due to reduced production.

As far as measuring the total market for compost products, market potential in horticulture is by far the largest (Slivka, 1992). The horticulture market has been considered by some to be the "dumping grounds" for composts which are not of the highest quality (Ozores-Hampton, 2017). It should be noted that farmers are usually in tune with their soils, often working with agronomists or university experts to determine fertilizer loading capacities and crop nutrient needs. Farmers are indeed the best land stewards on earth, compared to many urban counterparts. A few people feel that low-quality composts utilized by this market sector are acceptable, but if products containing inert materials (i.e., glass and plastics) are utilized, they will show up in the soil over time, causing permanent contamination. It seems that even the largest market of all will still demand quality composts.

Composts used in horticulture must be safe enough to avoid perma-nent contamination of soils with inert, heavy metals or other compounds (Ozores-Hampton, 2017). However, some composts may be applied in an immature state, which is less effective than fully mature composts (Ozores-Hampton et al., 1998a,b). It is wise not to plant immediately fol-lowing the application of immature composts due to nitrogen (N) immo-bilization which occurs to further break down free carbon.

Although many studies have been performed which illustrate the ben-efits of compost utilization on horticulture land, the market still refuses to pay high costs for these materials (Ozores et al., 2011, 2012). In general, normal farming practices can deplete over 50% of a native SOM over time (Lucas and Vitosh, 1978; Dawson, 2019). Also, losses of humus and other soil nutrients from erosion are significant in horticulture, but compost can help replenish these when added on a regular basis (Faucette et al., 2005).

12.1 Managing the organic matter (humus) pool by using compost in horticulture

Almost all compost used as soil amendment in the horticulture market focuses on incorporation or mulching (applied on top of the ground) field-grown plants, trees and shrubs (Rynk et al., 1992). Some conven-tional crops are not yet conducive to mulching due to lack of application equipment that makes these tasks quick and affordable. However, some specialty crops can be mulched and offer high dollar returns per acre, making the purchase of compost economically feasible (Rynk et al., 1992). Economic payback for funds invested in compost applications still need to be quantified with all crops in horticulture. This will take many years and many replications to develop evidence necessary to convince the "old school" farmers who still rely heavily on chemical fertilizer and pesticide applications. However, times are beginning to change.

Farmers adding large amounts of organic materials will eventually be rewarded. The Rodale Research Station in Emmaus, Pa., has estimated that it takes 6–8 years before a positive cash flow is generated from an organic-based farming program. This represents data from annual applications of small OM. However, if the system can be started with heavy compost applications, initial costs will increase but returns should be sooner than 6 years. For farms composting their own animal manure or accepting off-farm organic debris, starting fields with heavy compost applications may not necessarily involve expensive purchases of commercial grade compost. Organic matter additions improve drainage, buffering capac-ity, microbial activity, decrease soil bulk density and offer slow-release nutrients (Table 12.1) (Abdul-Baki et al., 1997a,b; McSorley, 1998; Sainju and Singh, 1997; Stivers-Young, 1998; Sullivan, 2003).

Table 12.1 General properties of organic matter (humus) and associated effects in the soil

Properties	Remarks	Effect on soil
Color	The typical dark color of many soils is caused by organic matter	Facilitates warming
Water retention	Organic matter (OM) can hold up to 20 times its weight in water	Helps prevent drying and shrinking
Combination with clay minerals	Joins soil particles into structural units called aggregates	Permits gas exchange, stabilizes structure and increases permeability
Chelation	Forms stable complexes with copper, manganese, zinc and other cations	Buffers the availability of trace elements to higher plants
Solubility in water	Insolubility of OM results partially from its association with clay	Limited OM is lost by leaching
pH relations	OM buffers soil pH in the slightly acid, neutral and alkaline range	Helps to maintain a uniform soil pH
Cation exchange capacity (CEC)	Total acidities of isolated fractions of humus range from 3,000 to 14,000 mmole/kg	Increases the CEC of soil; as much as 20%–70% of soil CEC can be caused by organic matter
Mineralization	Decomposition of OM yields CO_2, NH_4, PO_4^+ and SO_4^-	A source of nutrient elements for plant growth
Combination with organic molecules	Affects bioactivity, persistence and biodegradability of pesticides	Increases the application rate of pesticides for effective control

The degradation of OM is dependent upon soil aeration, moisture and temperature. At a soil temperature of 88°F, which is an ideal soil condition, OM is degraded faster than it can be naturally produced (Sachs, 1993a,b). In other words, hot humid climates may not be able to afford, depending on OM additions that accumulate from plant residue. Instead, a regular amendment program should be designed according to specific soil tests. Less than 1% of compost applications may become soil humus depending on conditions (Sachs, 1993a,b).

The advanced farmer confront high land uses to expect optimal yields year after year without any improvement to the land (Dawson, 2019). Due to the amount that the soil is tilled, OM losses through decay and erosion can be significant. In fact, on many parts of the nation with intensive crop production, soils have lost over 50% of their original OM (Robertson and Erichson, 1978; Swain, 1992). Future management of OM content in soils may become the farmer's most challenging task. A study in Michigan showed annual soil humus loss of 1,200 lb/acre on a soil originally testing 4.8% OM. Six years after 40 tons/acre of sawdust were applied, OM was only 0.6% higher (Lucas and Vitosh, 1978).

Remember that an acre furrow slice of soil (plow depth) weighs approximately 2,000,000 lb. An acre furrow slice with a 5% OM content would equate to 100,000 lb of humus. If 1,200 lb of that humus were lost annually, that equals 1.2% of soil humus lost yearly. Consider the equation in reverse if you must add 1,200 lb of actual humus to the soil yearly, to stay even. Since about 1% of most compost applications become humus, divide it by 0.01 (1,200/0.01 = 120,000 lb). Then take 120,000 lb of compost OM and divide by the average organic content in your product (i.e., 60%) 120,000 lb compost OM/0.6 = 200,000 lb compost. Now take the 200,000 lb compost and divide by average % dry weight of 200,000 lb compost/70% dry matter = 285,714 lb of compost. This is what would be needed to be applied to offset the 1.2% yearly loss of soil humus. At 142 tons/acre or about 284 cubic yards/acre, it is obvious that this would not economically be feasible.

It is vital for farmers to do all in their power to preserve the soil and existing humus which is so valuable. Under natural conditions, which include constant vegetative cover and no modern farmers, production and deletion of the annual average humus pool is about equal (Lucas and Vitosh, 1978). Soils without adequate management of organic applications (such as those under many commercial farming practices) have shown an organic depletion rate of about 1.2% per year (Lucas and Vitosh, 1978).

Some crops, such as cotton (*Gossypium barbadense* L.), have been grown successfully with 2–4 tons/acre or 4–8 cubic yards/acre of compost applied yearly. Although this represents a small application rate, the constant use year after year would create a large market capable of paying for compost applications from reducing cost savings (Goldstein, 1994).

Studies show that the regular application of raw horticulture materials, such as animal manure, does not readily change the OM content in soil over many years (Lucas and Vitosh, 1978). This may be due to the additional shrink which occurs during the decomposition process. Composting animal manures makes a more concentrated product organically, and so the impact on OM in soil after application is greater (Rosen and Bierman, 2005). Soil humus is lost on a regular basis to soil erosion and soil micro flora and is converted to carbon dioxide and water through natural processes. Research has shown that it takes 5–15 lb of fresh plant residue to produce 1 lb of humus while 10 tons of animal manure would only provide 0.5–2 tons OM (Hickman and Whitney, 1989).

12.2 Eroding soil in horticulture production

In 1978, it was estimated that the United States loses 3.6 billion metric tons of topsoil annually or about 14 tons/acre to erosion (Faucette et al., 2005; Lucas and Vitosh, 1978). This equates to about 1,000 lb/acre of soil humus. More recent studies indicate that the national average of soil lost from erosion is about 17.5 tons/acre. More recent information suggests we only have 60 years of farmable soil left on earth (Dawson, 2019). Water was our environmental focus in the 70s, air was the significant topic of the 80s and the soil will be our greatest challenge into and through the year 2000 and beyond.

The importance of OM in helping to reduce soil erosion cannot be overstated. For those thinking erosion is not that significant or that soil is easily replaced, consider this:

> Each year this country loses some five billion tons of topsoil to erosion, 80% of it is washed away, the rest blown away. This amounts to the equivalent of about 7 inches of topsoil from five million acres. Over the last 200 years, we have managed to lose roughly one third of the topsoil on the nations crop-lands and despite our best efforts at soil conserva-tion, we continue to lose topsoil at 10 times as fast as it is being formed.

(Swain, 1992; Dawson, 2019)

Since OM helps create aggregates when decomposing, it is logical to believe these aggregates would become less likely to be eroded over time (Gülser et al., 2015a,b). Compost applications offer the necessary OM needed to prevent the erosion of our nation's soils. The major dam-age from all erosion occurs through water erosion, it is interesting to further note effects from rainfall. Raindrops fall at roughly 20 mph, and someone has calculated that the kinetic energy generated by a 2-inch rainfall is enough to raise a 7-inch layer of topsoil 3 feet and on ground with 10% slope, 60% of the soil is thrown downslope, while only 40% splashes up (Swain, 1992).

Composts used in horticulture have shown a variety of benefits; one of the largest being reduction of erosion (Kashmanian, 1992). High inten-sity farming erodes valuable topsoil faster than it can accumulate natu-rally (Kashmanian, 1992). By adding compost on a regular basis, farmers can maintain healthy soils and stay profitable. Much research has been conducted over the last 20 years showing compost use can reduce ero-sion significantly compared to other erosion control alternatives (Faucette et al., 2008, 2009)

Horticulture exerts a tremendous downward, compacting force from all modern equipment used. Combatting this problem is the most challenging. For instance, a six-row planting and harvesting equipment line, about 15 feet wide, with 18 inches wide rear tractor wheels, will make enough wheel tracks during the normal season to cover every square inch of the field about twice (Voorhees, 1977). Compacted soils erode faster because they do not encourage infiltration, and as a result, have increased speeds of water moving on the surface during rainfall. Compost applications help reduce compaction and increase infiltration in heavy soils (Faucette et al., 2008, 2009).

Wheel traffic from heavy machinery, especially on slopes, creates automatic collection channels for surface water, increasing the erodibility of soils (Voorhees, 1977). When compost is used, the resulting soil springs back more, reducing wheel tracks and corresponding erosion potential.

Over the last four decades, failure to maintain effective soil conservation practices on horticulture land in the United States have resulted in a decline in soil productivity, accelerated soil erosion and nutrient runoff losses and decreased SOM (Dawson, 2019).

Although economic losses pertaining to lack of available moisture may be easy to calculate, the losses occurring from lack of infiltration even in irrigated areas may be harder to measure. Estimated annual losses from poor water use or infiltration can range from $165/acre (for irrigated pasture) to $988/acre for orchards in equivalent purchase power in 2020 (Boyle et al., 1989). In nursery crops where thousands of high dollar plants per acre are intensely managed, potential savings or losses are substantial.

12.3   Animal manure as an organic material in horticulture production

As enormous as the horticulture market is, the amount of compost from source-separated organics could never supply completely. Compost produced from animal manures in the horticulture market has potential to dwarf any other form of compost made from solid wastes. According to Environmental Protection Agency (EPA) estimates, 1.4 billion tons of animal manures are currently spread onto farmlands, almost exclusively in a non-composted or raw state (Rosen and Bierman, 2005). If all these organic residuals were composted, a vast amount, perhaps as much as a billion tons or cubic yards, of compost would be produced.

For farmers applying animal manure to support the maintenance of OM losses, the following information may be beneficial (Rosen and Bierman, 2005). One large Holstein "cow year" worth of animal manure is about 20 tons. While 20 tons of anything is a lot, when considering dairy manure, it translates into a small amount solid. If the approximately 5,200 lb of solid material in the 20 tons is applied over the surface of 1 acre and is mixed with 2,000,000 lb of soil present to a 6-inch depth, it would raise the SOM by about 0.3%. However, much of the animal manure will decompose during the year, so the net effect on SOM will be even less (Rynk, 1994; Magdoff and Weil, 2004).

The example above explains why many farm fields are not higher in OM, even after number of years of animal manure applications. A 75% decomposition rate in the above example would yield only 1,300 lb of OM, which happens to be about the same amount lost through natural processes in many soils (Magdoff and Weil, 2004). The message here is clear: if a farmer wants to increase the permanent OM of the soil, he should compost the animal manure first and apply the "concentrated" organic material to the field (Rosen and Bierman, 2005).

Composting of farm animal manures will increase in the future due to concerns from non-point source pollution currently facing agricultural waste disposal. In many of the intense horticulture areas, soils have more than 10 ppm of nitrates (NO_3) (Logsdon, 1993a). More and more farmers are considering composting as a management tool to prevent NO_3 and phosphorous (P) and reducing total volumes of material to be handled (Gagnon et al., 2012).

George Washington Carver, noted botanist, chemist and agriculturist who taught methods of soil improvement, recommended that farmers "compost materials and return them to the land to build up and maintain the virgin fertility of our soil." He advised that "a year-round compost pile is absolutely essential" (Ronald and Adamchak, 2008). Even with this testimonial, the horticulture sector remains the most underdeveloped market for compost. Composting of farm animal manure was estimated to be 60% less expensive than land application. With available application land in the region being scarce, due to the number of farms and the large amount of animal manure to dispose of, composting helps by reducing the total amount needing to be applied (Goldstein, 1994).

12.4   Contamination concerns by the horticulture market

A major concern in the horticulture market is compost quality (Ozores-Hampton, 2017). Even though many compost applications in the horticulture market are annual, the collective results of yearly applications can be permanent. This raises more concerns than most other markets because the horticulture market is linked to the food chain. Horticulture soils produce edible products for humans or animals consumption. Permanent contamination of farm soils with inert, heavy metals or other undesirable compounds may render the land unproductive in the future and raises all kinds of concern for liabilities. The United States simply cannot afford to use poor quality compost products on prime farmlands.

High-quality composts will only improve prime farmland when they are applied.

The source of potential contamination and the frequency are important to the public. The horticulture industry believes that composts with biosolids should be more carefully monitored than YW because of the potential for heavy metal or disease contamination (Goldstein and Steuteville, 1993; Jones, 1992). The fear of the unknown is often what guides public perceptions.

In a survey of California farmers, the highest concern rated among respondents was physical or chemical contamination (Grobe and Buchanan, 1993). A trial marketing program in Ohio, linking farmers with sawdust from a cabinet maker, failed because of the small pieces of formica residuals in the sawdust. Since the formica was white and was readily distinguishable, many farmers refused to set up trial loads to be used as soil amendment. Those involved in the survey identified the formica contamination as a major barrier to establishing a soil amendment program with the available sawdust.

The leading method to ensure that the potential danger for contamination is reduced is to compost only source-separated materials. Table 12.2 shows in column D the lowest of all in heavy metals due to separating the materials at the source. Although many of the long-term effects of regular compost applications with contaminants are not yet known, there is a logical relationship between cause and effect. There is current research indicating that chemical removal can be achieved at integrating compost into the operation, either as a perimeter measure or by tilling it in (Faucette et al., 2009).

12.5 Environmental benefits associated with composts applications

The simple task of playing "what if" before decisions are made can save all of us a lot of time and effort. For the part of life requiring change, the decision to change will result in things getting either better or worse. If we get positive results, we are encouraged to like change. If we get negative results, we are discouraged. By playing the "if-then" game ahead of time, we can better prepare ourselves for anticipated outcomes (Diacono and Montemurro, 2010).

1. Water holding capacity: mineral soils receiving 10–15 tons/acre of compost can generally be expected to increase in water holding capacity by 5%–10% (McConnell et al., 1993). This is obviously dependent upon initial soil conditions. Tester (1990) observed that up to twice as much water was available in a single application of compost 5 years later. The study was performed on sandy soils which respond favorably to water retention properties of compost.

Table 12.2 Heavy metal concentrations in municipal
solid waste (MSW)-derived compost

Metal	Processing method (mg/kg dry weight)			
	A	B	C	D
Zinc	1,700	800	520	230
Lead	800	700	420	160
Copper	600	270	100	50
Chromium	180	70	40	30
Nickel	110	35	25	10
Cadmium	7	2.5	1.8	1.0

A. Mixed household wastes are composted without preparation. The process takes approximately 12 months. After composting, the product is screened and inerts are removed.
B. The collected household wastes are separated into two frac-
tions. The material contains most of the easily degradable
organic material. Between 2½ and 5 months are needed for this
composting process.
C. The collected wastes are shredded and then processed,
resulting in a fraction to be composted. This fraction is free of
most inerts, such as glass and plastics.
D. Wastes are separated at the source. The organic components
are collected separately at households. All necessary steps are
taken to ensure that components containing heavy metals do
not enter the organic components.
Source: U. S. Environmental Protection Agency, 1994, 1995.

2. pH: an application rate of 10–20 tons/acre of a slightly alkaline
   compost usually increases pH by 0.5–1.0 in acid soils. While the
   pH interaction is often confusing, there is a time factor for reactions to occur that is somewhat immeasurable due to the variables
   involved, including moisture, temperature and soil type. Each
   individual soil should be tested prior to compost applications to
   determine the impact of the pH component in the total management plan.
3. Cation exchange capacity (CEC): the ability of a soil to hold nutrients
   for plant. It is a magnetic force which holds the nutrients. Similar
   to water holding capacity increases discussed earlier, CEC increases
   from compost applications of sandy soils are significant (McConnell
   et al, 1993). In fact, the increased CEC means less fertilizer leaching
   and greater nutrient availability to crops. This combination needs to
   be analyzed for each crop grown in each type of soil. Organic matter is a leading reason many soils have higher CECs, unless the clay
   content is high, and so understanding the correlation between CEC
   and OM additions is important to manage many growing systems.
   Table 12.3 identifies a direct positive relationship with increasing

Table 12.3 Effect of organic matter (OM) percent on cation exchange capacity (CEC)

	OM (%)					
pH	1	2	3	4	5	6
	CEC (meq/100 g soil)					
5	0.8	1.6	2.5	3.4	4.2	5.0
6	1.8	3.6	5.4	7.2	9.0	10.7
7	2.8	5.5	8.3	11	13.8	16.5

Source: Magdoff and Weil, 2004, calculated by equation.

OM and CEC values. Keep in mind that the further to the right of the table, the less dependent we are on fertilizer and other chemical nutrient inputs (Magdoff and Weil, 2004).

Therefore, these companies have no incentive to help the compost industry move further to the right. Marketing compost to horticulture will eventually affect sales in other markets (such as pesticides and fertilizers), probably negatively. If chemical companies do promote compost applications, they may be jeopardizing future sales. When other businesses are affected by compost applications, do not expect them to just stand there and watch their market slowly disappear. The relationships between chemical companies producing fertilizers and compost facilities really should be synergistic, not competitive. The higher CEC of compost will hold nutrients more tightly in the soil, reducing amounts of leachable compounds. Fertilizer will therefore be more effective, and fertilizer companies should receive less negative press from non-point source pollution concerns. Studies in fertilizer and compost interactions should be carefully analyzed. An important thought to consider when reviewing this research is: what is the agenda (motive) of the entity funding the research? If the funder is a compost facility, they want to sell all their product at as high a price as possible. If the funder is a fertilizer facility, they too want to maximize sales and profitability. From either source, research must be read with a suspecting eye and conclusions drawn based on a neutral point of view. It is often too easy to make data appear to be more significant than it is.

Beware of the promises made by some companies. It is not possible for miracle cures to occur from compost applications to soils. For instance, one leading national company reported increased profitability from compost applications even though yields were not as good as with conventional fertilizer program. Also, they reported that OM in the soils increased from 1.2% to 2.5% from the application of 2–3.5 tons/acre of compost. That was some compost! A 3.5-ton application of compost testing 70% OM would yield: 3.5 tons × 80% dry matter × 70% OM = 1.96 tons/acre of OM. Accordingly, 1.96 tons/acre of OM/2 million lb of soil = less than 0.001 increase in total OM of the parent soil.

4. Nitrogen availability: for some composts, within 1 year after application of 30 tons/acre of compost, about 150–200 lb on N is available (Kidder and O'Connor, 1993). The residual N content of compost applications is often another variable that depends on native soil type and climate. Table 12.4 shows typical release rates for temperate climates with adequate rainfall to drive chemical reactions, providing microflora ideal conditions.

Most farms, to rectify soil losses from erosion, land apply most of their animal manures. However, many of the animal manures currently being applied may contribute to non-point source pollution because they are more easily eroded and leached than products which are composted prior to application. Composted animal manures and farm wastes may help reduce non-point source pollution by converting nutrients into less leachable forms. By composting the animal manures prior to application, nutrients are locked up and rely on microorganisms to become available (Maynard, 1994). Figure 12.1 shows how the N release curve is similar to that of what plants require in order to maximize yield and, however, compared to a urea fertilizer, leach less which is more environmentally friendly (Tyler, 1996).

Evidence that compost can help reduce contamination of groundwater after fertilization is intensifying in the horticulture production. About 50% of the drinking water in the United States is supplied by groundwater, and when NO_3 reaches levels of 10 ppm or higher, danger exists. Nitrate leaching occurs after application of commercial fertilizers or non-composted animal manures (Maynard, 1994). For farms disposing of animal manure during cold weather, further regulations may restrict raw manure

Table 12.4 Total percent (%) availability of nitrogen (N) per year for three consecutive yearly compost applications containing 1% N

Application number	Year				
	1	2	3	4	5
1	25%	10%	10%	5%	5%
2	-	25%	10%	10%	5%
3	-	-	25%	10%	10%
Cumulative total	25%	35%	45%	25%	20%

Source: Tyler, 1994.

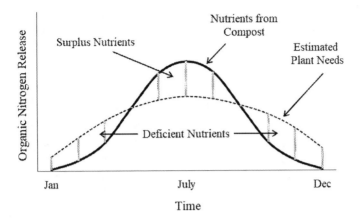

Figure 12.1 Total organic nitrogen (N) mineralization for temperate climates. Source: Tyler, 1996.

applications on water-saturated or frozen soil because of runoff potential into local rivers and lakes (Quaife, 1993). Anyone who has applied animal manure to sloped, frozen land just prior to a heavy rainfall knows the potential danger of contaminating nearby watersheds. Ironically, composted animal manures do not pose as great a danger, and so they offer great flexibility for proper farm animal manure disposal and increased window of time for applications. Maynard (1994) found that applications of 0.5 and 1 inches of composts each year for 3 successive years resulted in a slight increased NO_3 concentration in groundwater. Although all treatments of compost plots remained below 10 ppm NO_3 in groundwater, plots with commercial fertilizer nearly reached 15 ppm (Maynard, 1994). This suggests that a cumulative effect of fertility from composts is first released and then becomes leachable in the soil profile. Since the maximum concentrations in the compost-treated plots was just under 10 ppm, Maynard's (1994) study suggests three yearly applications may be the limit for compost applications based on NO_3 measurements alone. It has also been demonstrated that more mature composts have a higher concentration of organic N in relation to total N. Similarly, Tyler (1994) also suggests that the third year is the time when additive effects from yearly compost applications may become excessive (Tyler, 1994). These additive effects of available N levels in soils, provided by compost applications, are shown in Table 12.4.

For farmers under pressure from regulations affecting non-point source pollution, compost applications may produce significant dividends. The focus is simply shifted from economics associated with production to economics associated with insurance and potential liability from pollution. With normal fertility and animal manure management programs, depending on the soil and rainfall, much of the fertilizer and nutrients applied can leach into the groundwater (Diacono and Montemurro, 2010). Compost applications may help reduce this leaching and therefore potential contamination of surface wells (Grobe and Buchanan, 1993). Future direct land application standards for animal manure may be based on P limitations, especially in areas where P have been elevated in the ecosystems (Ozores-Hampton, 2017). Since farmers have traditionally applied nutrients based mainly on N requirements, regulations based on P requirements will be challenging to meet because of traditional production practices. Table 12.5 is a good example of how application rates of animal manure may differ for the land required, depending on whether N or P is used for application guidelines. The comparison of how much land it takes to apply animal manure to a corn crop, in an agronomically correct, based on both N and P requirements. These are rules only, and application rates may vary widely according to crop rotation, yields, animal manure and soil tests (Cubbage, 1994).

The number of acres needed for animal manure applications on a 500-cow dairy if P is used as the limiting factor, 746 acres are needed for continuous corn compared to 218 acres if N is used as the limiting factor. The additional acreage required, almost three times the amount required for traditional N limitations, is a challenge for any farmer to meet overnight. Therefore, if the P standard becomes accepted in all states (and it is already well on its way), farmers will have more incentive than ever to compost the animal manure to avoid purchasing additional land for animal manure application.

The Pennsylvania (USA) Nutrient Management Act was passed in 1993 in farm nutrient management. Large animal operations must develop a plan for animal manure (nutrients), and if residual animal manure is identified, the plan must state plans for use (Riggle, 1994). Many farms

Table 12.5 Comparison of land base needed for animal manure applications based on nitrogen (N) and phosphorous (P)

Number of cows	N (lbs)	P_2O_5 (lbs)	Acres required for continuous corn		Acres required for corn after and soybeans	
			N	P_2O_5	N	P_2O_5
50	3,277	3,358	22	75	30	75
100	6,554	6,716	44	149	60	149
200	13,107	13,432	87	298	119	298
500	32,768	33,580	218	746	298	746
1,000	65,536	67,160	437	1,492	596	1,492

Source: Cubbage, 1994.

have chosen composting as a method to reduce leaching and make animal manure more valuable to crops (Riggle, 1994). When animal manure is stockpiled to reduce the non-point source pollution concern and applied only when good weather prevails, it can lose up to 50% of initial N values, reducing the agronomic value to the farmer (Logsdon, 1993b). However, composting the animal manure prior to application would reduce losses of nutrients and decrease or prevent leaching.

5. Phosphorous: this is a major concern in non-point source pollution issues. State officials in Florida cited P from dairy pasture runoff as the cause of increased algae blooms in Lake Okeechobee, FL (Cubbage, 1994). This led to further regulations and the adoption of an animal manure application standard based on P content. Phosphorous in dairy manures can range from 6 to 28 lb per 1,000 gal of animal manure, which means farmers may not be able to apply as much manure to their lands as they have in the past (Rynk, 1994; Cubbage, 1994). The implications are that farmers need more land just to comply with new P guidelines. Since animal manures are high in moisture content and their nutrients easily leached, it makes sense for these "wastes" to be composted first. Composting will alter the state of many of the nutrients, making it more difficult for them to leach. More importantly, when compost is applied, it is in a drier form that is less prone to runoff during heavy rains.

In the Great Lakes region, there is more concern with P than other areas due to many issues associated with surface waters (Cubbage, 1994). Excess P is extremely stimulating to aquatic plant life, often growing up to 500 lb of OM per lb of P (Cubbage, 1994). There is a misconception about how P enters surface waters. Unlike N, which is mobile in certain forms and can leach readily through soil profiles, P is not as mobile. Phosphorous tends to stay attached to soil particles and move with these particles into runoff entering lakes and streams (Cubbage, 1994; Gagnon et al., 2012). Another real benefit of using compost on farms and from composting animal manures will be to help reduce soil erosion and in turn reduce P pollution.

12.6 Economics of compost application

The low value placed on composts entering horticulture markets may require a future "disposal fee." Unfortunately, farmers may be fooled into thinking short term. Any composter that pays a fee to take their product proves that the compost facility owners know their product is valueless. Off-specifications compost may be available and appear to be a "deal" for farmers, but long-term physical or chemical contaminants will prove to become the farmers' worst nightmare. Future legislation may also hold the farmer responsible for contaminating the land and may require expensive remediation procedures if severe.

A survey in California revealed the compost marketing challenge to horticulture as twofold: increasing total number of compost users as well as total volume of each application. As reported, these applications would increase by 50%–150% if price and quality were within acceptable levels (Grobe and Buchanan, 1993). It is obvious from this survey that a key factor limiting horticulture compost applications is price. What would happen if compost were free? what if farmers were paid to take compost?

Further analysis of the higher dollar markets, volume markets and their relationship specifically to the horticulture market draws severe concern. If the horticulture market is rendered damaged from the use of inferior compost, the next logical solution would be to dump the low-quality products into other higher dollar markets at bargain prices. The permanent market damage would eventually result in destruction of the entire market. This damage, however, would take longer to see in the higher dollar markets because many of them are based on one-time applications. Once the economics are affected in the higher dollar market, an internal collapse would occur, and an entire industry could be lost.

Monetizing the benefits of compost as mentioned before is challenging due to several factors; however, when using compost, we must add all the collective benefits for the entire period they are provided. There is limited data on this subject to date.

12.7   Equipment required for compost application

Understanding revenues per acre for each crop grown in the horticulture market may help to identify which crops can "pay for themselves" by increased yields, thus covering costs of compost additions. For fruits and vegetables, the return is considerably higher than traditional row crops of corn and soybeans. Typical compost applications range from 2 to 20 tons/acre with the average being between 5 and 10 tons/acre (Grobe and Buchanan, 1993). This amount is still relatively low compared to accepted landscape applications but promises to increase in the future when more evidence is available to document increased yields or when the price of compost is no longer a factor.

The costs of the compost alone are not the only additional roadblock to the horticulture market development. According to growers using horticulture animal manures as an amendment, 79% indicated that they rely on custom application services to complete the reuse program (Grobe and Buchanan, 1993). With the number of variables to manage at any farm, the thought of having to purchase compost and arrange trucking and spreading may seem more problematic than it is worth. Especially when evidence of increased economic performance of the crop may not be demonstrable.

For specialized application companies to grow by offering custom compost applications, there are still limitations. Most professional compost application companies can spread about 300–500 tons/day (Grobe and Buchanan, 1993). Using 10 tons/acre as an average, the company would be limited to 50 acres per day per crew (loader, spreader and hauler). For intense horticulture areas where applications of fertilizer on 1,000 acres is common, the 50-acre maximum would not satisfy the demands of large farms. Perhaps that is the main reason most commercial fertilizer dealers do not currently have compost application programs.

12.8   Horticulture industry using compost

For horticulture crops "paying for compost applications," that is, where demonstrable economic benefits are known, applications of as much 3-6 inches of compost are incorporated with an additional 4-inch layer as mulch (Grobe and Buchanan, 1993). On a large scale, this would result in a maximum application rate of 675 tons/acre of compost. For these special applications, it is necessary to have controlled conditions which are available in greenhouse cropping systems, enabling runoff water to be captured and recycled to avoid possible NO_3 leaching.

In a few cases, horticulture (volume) producers of compost market excess materials to dollar markets, creating added farm income (Rynk et al., 1992). While this is not common, the possibility exists for this activity to increase, especially for farms located near urban-rural fringes. Although documented evidence of economic benefits to farmers are scarce, general benefits from compost applications are starting to escalate (Stentiford and Sanchez-Monedero, 2016). A theory on how this occurs is at least physically explainable. Since compost helps increase drainage for most heavy soils, the reduction of puddling that occurs during and after a rain may help reduce *phytophthora*, a soil disease (Logsdon, 1993b). This certainly will be a more permanent solution than spraying chemical pesticides every time a disease outbreak occurs. More and more farmers are looking to utilize more permanent solutions rather than the traditional approach using chemicals pesticides.

There may be only three basic methods of natural disease suppression from using compost. Specific suppression relates to the actions of specific microbes, general suppression occurs through competition for nutrients and chemical suppression may result from production of fungus inhibiting by-products of microbes in the soil (Hoitink et al., 2001; Mecklenburg, 1993). The types of suppression are usually not known in the field rather than that the disease incidence has decreased.

As more studies by the scientific community is been reported, it becomes clear that there may be general disease suppression with many horticulture crops (Hoitink et al., 2001). Instead of expecting scientists to study every crop, disease, compost and soil series, it may be faster for the marketer to coordinate small-scale field trials with crops within market areas of available quality compost products. By doing this, marketers and end-users will not be confused about what to promise or what to expect. For example, soil disease suppression in peppers (*Capsicum annuum* L.) when compost was applied as a mulch may be vitally important depending on the management practices of each farmer. For instance, farmers in dry climates may apply compost pre or post planting while wetter climates may have no choice but to wait until the plants are 6 inches high, after the soil has sufficiently dried (Logsdon, 1993a and b). It will also require less labor to mulch a field post planting (pre-germination) if the plants are able to grow through the compost mulch layer.

Additional benefits besides soil disease suppression may be in reducing the fertilizer applications by half without reducing yields (Hoitink et al., 2001; Logsdon, 1993b). This may be the result of plants establishing themselves in looser, moister, more nutrient dense soil; they establish larger root masses which improve them to uptake nutrients more efficiently. The reduction in fertility program by 33% can still produce the similar tomato (*Solanum lycopersicum* L.) yield in compost-amended plots (Logsdon, 1993a). Other studies suggested the reduction in diseases such as mosaic virus, fusarium crown rot and rhizoctonia root rot (Hoitink et al., 2001; Logsdon, 1993b).

12.9   A peek into the compost future?

The entire horticulture industry, an especially important market, has become extremely competitive. The industry also has developed to such a finite level that most farmers know what return they must obtain prior planting a crop. In other words, if a farmer has a choice between two fertilizers, the more expensive option must deliver a proportionately greater harvest to offset initially higher costs. Likewise, if a farmer uses compost, he must understand the economic return necessary to offset the expense. Many farmers have been short sighted when it comes to budgeting for next years' nutrient requirements because the short-term decisions are made on dollars alone. The economics might improve greatly if the farmer receives compost for free. However, if the free compost has a lot of contaminants or poor quality and is used on a yearly basis, it will eventually render the farmers' land useless for horticulture purposes. On higher value crops that are intensely grown, such as strawberries, tomatoes, melons (*Cucumis melo* L.) and peppers, compost applications are easier to justify due to the economic return per acre. The return per acre from these crops is indeed higher than traditional row crops but is not as easy to develop economies of scale due to the amount of labor required for harvest. On the other hand, the vast acreage planted with corn and soybeans allows a farmer to take advantage of economies of scale and mechanization. Mechanical pickers for strawberries are only affordable by the largest of strawberry farms.

It may be concluded that due to the disparity in economic return per acre, it is more economically logical for marketers to address horticulture enterprises capable of generating a higher per acre return. Studies are being conducted to determine yield differences with these various crops grown in compost-amended soils. Early results are positive, but more work in this area is needed, with a variety of available composts. Trends in organic farming and demand for organically produced vegetables are increasing consistently over the last 40 years. The link between compost and organic farming is simple in theory since limited economic proof is available to support or encourage this link.

The number of benefits associated with the addition of compost to horticulture fields appears to far outweigh the extra effort and costs associated. However, an educational system is needed to light the way for market development in large horticulture areas. By exploring compost products as natural resources that can be utilized to assist offset losses of soil erosion, we cannot forget that soils that receive compost applications are one of our largest natural resources. Soil permanent damage or contamination to these vast resources should lead quality control planners to stand strong on high-quality standards to ensure adequate land is available for application indefinitely to future generations.

References

Abdul-Baki, A.A., J.R.Teasdale, and R. Korcak. 1997a. Nitrogen requirement of fresh-market tomatoes on hairy vetch and black polyethylene mulch. *HortScience* 32:217–221.

Abdul-Baki, A.A., R.D. Morse, T.E. Devine, and J.R.Teasdale. 1997b. Broccoli production in forage soybean and foxtail millet cover crop mulches. *HortScience* 32:836–839.

Boyle, M., W.T. Frankenberger, and L.H. Stolzy. 1989. The influence of organic matter on soil aggregation and water infiltration. *J. Prod. Agric.* 2(4):290–298.

Cubbage, S. 1994. Look out for phosphorus. *Dairy Herd Manage.* 31:42–48.

Dawson, R. 2019. The need to grow. Earth Conscious Films. Creative Visions Foundation. 28 May 2010. https://www.earthconsciouslife.org/theneedtogrow.

Diacono, M., and F. Montemurro. 2010. Long-term effects of organic amendments on soil fertility. A review. *Agron. Sustain. Dev.* 30(2):401–422.

Faucette, L.B., J. Govermo, R. Tyler, G. Giglet, R.F. Jordan, and B.G. Lockaby. 2009. Performance of compost filter socks and conventional sediment control barriers used for perimeter control on construction sites. *Soil Water Conserv. Soc.* 64(1):81–88.

Faucette, L.B., C.F. Jordan, L.M. Risse, M. Cabrera, D.C. Coleman, and L.T. West. 2005. Evaluation of storm water from compost and conventional erosion control practices in construction activities. *J. Soil Water Conserv.* 60(6):288–297.

Faucette, L.B., K.A. Sefton, A.M. Sadeghi, and R.A. Rowland. 2008. Sediment and phosphorus removal from simulated storm runoff with compost filter socks and silt fence. *J. Soil Water Conserv.* 6(4):257–264.

Gagnon, B., I. Demers, N. Ziadi, M.H. Chantigny, L.E. Parent, T.A. Forge, F.J. Larney, and K.E. Buckley. 2012. Forms of phosphorus in composts and in compost-amended soils following incubation. *J. Can. Soil Sci.* 92:711–721.

Goldstein, J. 1994. Poultry processor favors composting. *BioCycle* 35(2):26.

Goldstein, N., and R. Steuteville. 1993. Biosolids composting makes healthy progress. *BioCycle* 34(12):48–57.

Grobe, K., and M. Buchanan. 1993. Agricultural markets for yard waste compost. *BioCycle* 34(9):33–36.

Gülser, C., F. Candemir, Y. Kanel, and S. Demirkaya. 2015a. Effect of manure on organic carbon content and fractal dimensions of aggregates. *Eurasian J. Soil Sci.* 4:1–5.

Gülser, C., R. Kızılkay, T. Askın, and I. Ekberli. 2015b. Changes in soil quality by compost and hazelnut husk applications in a hazelnut orchard. *J. Compost Sci. Util.* 23:135–141.

Hickman, J.S. and D.A. Whitney. 1989. Soil conditioners. *North Central Regional Extension Publication* 295:1–4.

Hoitink, H.A.J., Krause, M.S, and Han, D.Y. 2001. Spectrum and mechanisms of plant disease control with composts. In *Compost utilization in horticultural cropping systems*, edited by Stofella, P.J., Kahnn, B.A., 263–274. Boca Raton, FL: Lewis Publishers.

Jones, P. 1992. *Market Status Report Compost.* California Integrated Waste Management Board Staff Report, August, pp. 11–15.

Kashmanian, R. 1992. Composting and agricultural converge. *BioCycle* 33(9):38–40.

Kidder, G., and G.A. O'Connor. 1993. Applying non-hazardous waste to land: II. Overview of EPA's 1993 sewage sludge use and disposal rule. Fla. Coop. Ext. Serv. Univ. of Florida, Gainesville, FL.

Logsdon, G. 1993a. Turnaround in the poultry industry. *BioCycle* 34(2):60–63.

Logsdon, G. 1993b. Using compost for plant disease control. *BioCycle* 34(10):33–36.

Lucas, R.E., and M.L. Vitosh. 1978. *Soil Organic Matter Dynamics.* Michigan State University Agricultural Experiment Station, East Lansing. Research Report, November, pp. 2–11.

Magdoff, F., and R.R. Weil. 2004. Soil organic matter management strategies. In *Soil Organic Matter in Sustainable Agriculture*, edited by F. Magdoff and R.R. Weil, 45–65. Boca Raton, FL: CRC Press.

Maynard, A.A. 1994. Effect of annual amendments of compost on nitrate leaching in nursery stock. *Compost Sci. Util.* 2(3):54–55.

McConnell, D.B., A. Shiralipour, and W.H. Smith. 1993. Compost application improves soil properties. *BioCycle* 34(4):61–63.

McSorley, R. 1998. Alternative practices for managing plant-parasitic nematodes. *Am. J. Alter. Agr.* 13:98–104.

Mecklenburg, R.A. 1993. Compost cues. *Am. Nur.* 2(78):63–71.

Ozores-Hampton, M. 2017. Guidelines for assessing compost quality for safe and effective utilization in vegetable production. *HortTechnology* 27:150.

Ozores-Hampton, M., T.A. Obreza, and G. Hochmuth. 1998a. Using composted wastes on Florida vegetables crops. *HortTechnology* 8:130–137.

Ozores-Hampton, M., T.A. Obreza, and P.J. Stoffella. 1998b. Immature compost used for biological weed control. *Citrus and Vegetable Magazine.* March 12–14.

Ozores-Hampton, M., P. Roberts, and P.A. Stansly. 2012. Organic pepper production. In *Peppers: Botany, Production and Uses*, edited by V. Russo, 165–175.
Oxfordshire, UK: CABI.

Quaife, T. 1993. Stink over manure will just get worse. *Dairy Herd Management.*
Nov., pp. 7.

Ozores-Hampton, M.P., P.A. Stansly, and T.P. Salame. 2011. Soil chemical, biological and physical properties of a sandy soil subjected to long-term organic
amendments. *J. Sustain. Agr.* 353:243–259.

Riggle, D. 1994. Publicly supported composting research projects in the U.S.
Compost Sci. Util. 2(1):32–41.

Robertson, L.S., and A.E. Erickson. 1978. Soil compaction. *Crops Soils Mag.*
30(6):8–10.

Ronald, P.C., and R.W. Adamchak. 2008. *Tomorrow's Table: Organic Farming,
Genetics, and the Future of Food*, pp. 232. Oxford University Press.

Rosen, C.J and P.M. Bierman. 2005. Using manure and compost as nutrient sources
for fruit and vegetable crops. *Univ. Minnesota Ext. Serv.* 28 May 2020. https://
conservancy.umn.edu/handle/11299/200639.

Rynk, R. 1994. Status of dairy manure composting in North America. *Compost Sci.
Util.* 2(1):20–26.

Rynk, R., M. van de Kamp, G.B. Willson, M.E. Singley, and T.L. Richard. 1992.
On-Farm Composting Handbook. New York: Northeast Regional Agricultural
Engineering Service.

Sachs, P.D. 1993a. Nature provides best compost formula. *Bioturf News. Landscape
Management*, December, pp. 9–30.

Sachs, P.D. 1993b. Preserve soil's organic matter with a balanced compost mix.
Bioturf News, Landscape Management, October, pp. 28–29.

Sainju, U.M., and B.P. Singh. 1997. Winter cover crops for sustainable agricultural systems: influence on soil properties, water quality, and crop yields.
HortScience 32:21–28.

Slivka, D.C., T.A. McClure, A.R. Buhr, and R. Albrecht. 1992. *Potential U.S.
Applications for Compost*. Application Study Commissioned by The Proctor &
Gamble company for the Solid Waste Composting Council.

Stentiford, E. and Sanchez-Monedero, M.A. 2016. Past, present and future of composting research. *Acta Hortic.* 1146:1–10. doi: 10.17660/ActaHortic.2016.1146.1.

Stivers-Young, L. 1998. Growth, nitrogen accumulation, and weed suppression by fall cover crops following early harvest of vegetables. *HortScience*
33:60–63.

Stoffella, P.J., Z.L. He, S.B. Wilson, M. Ozores-Hampton, and N.E. Roe. 2014.
Utilization of Composted Organic Wastes in Vegetable Production Systems.
Food & Fertilizer Technology Center, Technical Bulletins. Taipei, Taiwan. 3
May 2020. http://www.agnet.org/htmlarea_file/library/20110808105418/
tb147.pdf.

Sullivan, P. 2003. Overview of cover crops and green manures. Appropriate technology transfer for rural areas. *Natl. Sustainable Agr.* Information Ctr.

Swain, R.B. 1992. Nutrient recycling. *Horticulture.* June/July, 27–33.

Tester, C.F. 1990. Organic amendment effects on physical and chemical properties
of sandy soil. *Soil Sci. Soc. Am. J.* 54:827–831.

Tyler, R. 1994. Fine-tuning compost markets. *BioCycle* 35(8):41–48.

Tyler, R. 1996. *Winning the Organic Game – The Compost Marketers Handbook.*
Alexandria, VA: ASHS Press.

U. S. Environmental Protection Agency (USEPA). 1994. *A Plain English Guide to the EPA Part 503 Biosolids Rule.* EPA832-R-93-003. Sept. Washington, DC.

U. S. Environmental Protection Agency (USEPA). 1995. *A Guide to the Biosolids Risk Assessments for the EPA Part 503 Rule.* EPA832-B-93-005. Sept. Washington, DC.

Voorhees, W.B. 1977. Soil compaction: our newest natural resource. *Crops Soils Mag.* 29(5):13–15.

Index

Printed in the United States
By Bookmasters